Multieffects for Musicians

by Craig Anderton

Multieffects for Musicians

by Craig Anderton

Amsco Publications
New York • London • Paris • Sydney

Cover photo courtesy of Eventide Inc.
Interior design and layout by Len Vogler
Interior illustrations by Craig Anderton and Len Vogler

Order No. AM 91245
US International Standard Book Number: 0.8256.1447.3
UK International Standard Book Number: 0.7119.3488.6

Exclusive Distributors:
Music Sales Corporation
257 Park Avenue South, New York, NY 10003 USA
Music Sales Limited
8/9 Frith Street, London W1V 5TZ England
Music Sales Pty. Limited
120 Rothschild Street, Rosebery, Sydney, NSW 2018, Australia

Printed in the United States of America by
Vicks Lithograph and Printing Corporation

Preface

Whether you want shimmering guitars, garage band drums, in-your-face vocals or earth-trembling bass, multieffects are an exceptional tool for self-expression because they combine sophisticated signal processing with computerized "total recall" of your favorite sounds. Furthermore, most processors include MIDI (Musical Instrument Digital Interface), which allows for further shaping your timbre—even to the point of totally automating all effects changes, whether live or in the studio. Unfortunately, because musicians are frustrated by complicated manuals and operating systems, few of them take full advantage of today's multieffects.

This book was written to show you the potential of this new generation of signal processors. It's been said that most people use only 5% to 10% of their brain's capacity; the same is true of multieffects. If you've held off getting into these devices because you're intimidated, don't be. If you can play music—which is a *very* complex process—you can learn about multieffects. Hopefully this book will shorten your personal learning curve, and give you a mastery of signal processing that lets you truly express your musical identity.

Why be normal? With a little effort, you can create your own "signature sound" to help you get ahead in today's ultra-competitive music scene. Go for it, and have fun in the process.

Craig Anderton

Acknowledgements

Several manufacturers have been kind enough to loan me multieffects for research into this book, but I'd particularly like to thank Peavey, Alesis, Ensoniq, Digitech, Symetrix, ART, Rane, and ADA for their willingness to educate me about the fine points of what some of these devices can do.

This book was started while in Belize (Central America), far away from faxes and phones, where I had a chance to do some serious writing without the distractions of everyday life. I'd like to extend my thanks to Lynne Werney for making this possible. It was finished during my stay as a teacher at the National Guitar Summer Workshop in Connecticut and Toronto, where probing questions from several students helped give me a better of idea of what information to include.

Finally, this book is dedicated to the people of Dangriga, Belize, who gave me some new and welcome insights into what music is all about. I'll be back...

Contents

CHAPTER 1
Shopping Tips for Multieffects

Musicians often ask me for advice about which multieffects to buy, as if there was one "perfect" effects box that would fill all their needs. Yet every musician has different desires, and a device that might be perfect in one context might be useless in another.

The secret to choosing the right multieffects is to define *your* needs as specifically as possible. It's like buying a car: do you want a vehicle that takes corners fast, or can carry all your band's gear? One with the lowest maintenance record, or one that's the least expensive to repair if something does go wrong? Will you be driving mostly on mountain roads in California, or highways in Texas? How important is fuel economy? Stick or automatic transmission? And once you get these basic questions out of the way, there's still the matter of the paint color, whether you want air conditioning, if you're willing to install a stereo system yourself and save money compared to getting one from the factory, and so on.

Here's an example of how one's requirements dictate a particular processor. When playing guitar, I try not to depend on a particular amplifier to get "my" sound. So, I use lots of signal processing before the guitar hits the amp input, and a clean amp. This gives me the same basic sound whether I plug into a mixing console or a PA system. What I require in a multieffects is great distortion and equalization (tone control) because that's what allows for simulating the sound of various amps. Also, since I plug my guitar directly into the multieffects, I want to make sure it's designed to handle low-level guitar signals.

A guitarist who relies on a particular amp's distortion characteristics would generally prefer a signal processor that delivers superb clean effects such as chorus, delay, and reverb. Since the multieffects would probably insert in an amp's effects loop (in other words, after the preamp but before the power amp; see Chapter 6 for more information on interfacing with effects loops), being able to handle low-level signals isn't as much of a priority.

If you're mostly interested in processing voice, you probably couldn't care less about distortion but will want superb reverb and pitch transposition. For me, reverb is the least important part of a processor: when playing live I don't need reverb, and when I'm in the studio, I'd rather go through a big-bucks dedicated reverb unit anyway. A keyboard player whose MIDI synthesizer lacks on-board EQ will want a multieffects with excellent EQ, along with a comprehensive MIDI implementation to take advantage of the MIDI control signals that emanate from the synthesizer.

Another consideration is whether you'll program your own sounds. If not, then choose a device with the best-sounding factory presets. Otherwise, look for a unit that's easy to program (not all multieffects are equally user-friendly) and has a well-written manual.

Get the picture? As you soon as you start shopping for a multieffects, your first task is to decide on your priorities. The next step is to visit several music stores and find out what's available. Remember that different stores specialize in different lines, so Store A might push one brand and Store B, another brand. Don't be swayed by sales talk or product endorsements from superstars; let your ears and requirements be the judge.

While you're in the store, sit down with each multieffects that interests you and go through all the factory presets. These try to give a wide range of effects to show the unit's capabilities, so if few of the programs excite you, it's probably time to move on (although the problem might be that the device was programmed by someone with musical tastes very different from yours). Narrow your search down to the units whose basic sound pleases you the most, then peruse the details. Accept the fact that although you'll never find the perfect multieffects, if you're willing to take some time and do some research you *can* find the one that's right for you.

Adventures in Specs-Land

Some particular features (as described in the following list) might tilt your decision toward one unit over another. While no device includes all these features, some come pretty close. If you're not familiar with some of the terms, check the glossary or the relevant sections in other chapters.

Input/Output Features

- **Ability to accommodate low and high level inputs.** Although all modern multieffects work with high-level output devices such as drum machines, tape tracks, and synthesizers, some also have the proper impedance and gain structure to work with low-level signals such as guitars and microphones. For guitars, unless you're using active pickups or an on-board preamp, make sure the multieffects' spec sheet shows an input impedance of 100kΩ or more. Anything less could reduce your instrument's high frequencies and level.

Also note that some multieffects are not designed to accept a straight, unamplified mic output. To check this out, plug in a microphone and make sure you that it can give enough level to light the "overload" LED if the multieffects input level control is all the way up.

- **XLR connector balanced outputs.** Although not essential for project studio and live applications, when interfacing with professional studios XLR balanced connectors provide compatibility with pro-level mixers.

- **Paralleled input and output jacks.** Multieffects with paralleled input and output jacks on both the front and rear panels are very convenient. Use the rear panel connections for permanent installations, and the front panel ones if you need to patch quickly into a system.

- **Headphone jack.** Why not get a headphone amp thrown in for free? This lets you audition patches quickly, practice at night without disturbing the neighbors, and tune up on stage in an emergency without anyone else having to hear it.

Sound Quality

- **Smooth changes from one program to another.** When you change from one programmed sound to another (either from the front panel or via MIDI program change commands), does the device burp and make undesirable sounds, or does it switch seamlessly? Check this out with complex time-based effects such as delay, tapped delay, pitch transposition, reverb, and so on, as these are most prone to glitching when switched.

- **Hiss-free distortion.** Dial up a program that uses distortion and determine whether a noise gate is in the processing chain. If it is, bypass the gate to see how much noise the distortion *really* generates.

- **Lack of bypass switch pops.** Hit the bypass switch with no signal going through the unit; you should not hear any clicks or pops. Also try bypassing in the middle of a sustained chord. The sound should change quickly, but not generate any clicks. You also want to avoid units where there is a perceptible delay going in or out of bypass mode.

Architecture

- **Hardware expansion options.** Some multieffects let you augment their basic capabilities via additional memory modules or cartridges; these let you increase the number of programs or effects. While most signal processors are not easily expandable, those that are help forestall obsolescence.

- **Performance tradeoffs due to memory limitations.** A multieffects may be advertised as having a zillion different effects, but in general the more sophisticated the effects you want to use, the fewer you can use at the same time. For example, it may be impossible to use pitch transposition, reverb, and distortion simultaneously, while compression, EQ, and delay are no problem.

- **Effects routing options.** Some multieffects allow only series connections of effects, while others allow serial, parallel, series/parallel, and other options. Also, most multieffects provide certain fixed effects algorithms (combinations of effects), while others let you make up your own algorithms. The latter approach, though rarer and more costly, is more flexible.

- **Stereo operation.** Since many instruments provide a mono output, most multieffects synthesize a stereo field for effects such as reverb, chorus, flanging, etc. However, for studio or electronic instrument applications, look for true stereo operation that provides independent processing for the right and left channels. Some multieffects even let you apply different effects to the two channels (dual mono operation)—for example, chorused guitar through the right channel and reverberated vocals through the left channel. The importance of true stereo operation, or dual mono operation, depends mostly on your application.

- **Memory backup method.** A multieffects should offer an easy, convenient way to backup programs, such as MIDI system exclusive storage or removable memory cartridges.

Hardware Features

- **Reliable buttons.** Press the buttons. Does each press have a positive feel, or do the buttons act intermittently? If the latter, they will only get worse with time.

- **Cool operation.** Leave a unit on for 30 minutes or so, then feel the case. If it's cool or just warm to the touch, that's a sign of a conservatively rated power supply with adequate heat sinking, both of which promote more reliable long-term operation.

- **Internal power supply.** You may want to avoid effects that have external power supplies (often referred to as "wall warts"). External supplies hog space on barrier strips, and if the supply is a custom design that provides an off-the-wall voltage (sorry about the pun), you're in trouble if it gets lost or broken. Furthermore, external power supply connectors often use flimsy miniconnectors that don't seem particularly roadworthy.

  However, there is another side to this story: transformers generate hum, so mounting one inside a box requires shielding to keep it from interacting with other components. Furthermore, if you have several rack units mounted on top of one another in a rack, the transformer inside one box can interact with a poorly-shielded box installed above or below it.

  Incidentally, if you run out of space on your barrier strip because of wall warts, there's a simple project you can do to reclaim some of those spaces (see sidebar, "Wall Wart Tamer").

- **Clear, readable display.** LCD displays should be backlit, but that's not all—you should also be able to see the display from a wide variety of angles. Some designs are inherently easy to see; others include contrast or viewing angle controls to compensate for different viewing angles.

- **No mechanical noise.** Turn off your amp or monitoring system and listen to whether the unit emits any loud noises—we're not referring to sounds that show up in the audio output, but the sound of the unit itself. Buzzes or hum may be the symptom of a low-quality transformer (or loose screws, which tells you something about the workmanship). While you're at it, shake the unit gently a few times to make sure there's nothing loose rolling around inside.

- **Companion foot controller.** A footswitch designed to work with a specific multieffects may offer features that aren't available with a more "generic" MIDI footswitch.

- **Three-conductor line cord.** The third ground pin offers a great deal of protection from electrical shock (even though it can increase hum if you're not aware of proper grounding procedures). Also, detachable power cords are preferable; when packing up, you don't necessarily want to have a cord trailing around. The downside is that it's easier to lose a detachable cord, so always keep a spare in your toolkit.

- **110/220V switch.** Maybe your club band will never tour Europe but maybe it will, in which case having this switch certainly beats buying a transformer/adapter.

- **Externally-mounted fuse.** Fuses can blow for a variety of reasons, and having to open up a box when this happens is inconvenient. However, a fuse post sticking out of the back of a unit does make it more vulnerable to damage if something smashes into it.

- **Rack-mount capability for half-rack units.** Half-rack units are cute and cost-effective, but make sure they can also mount in a standard rack if necessary.

- **As much user-editable program memory as possible.** 128 programs may seem like a lot—and it is—but just as with computers, you can never have too much memory. Also make sure you like the factory presets; they're usually in there permanently.

- **The ability to name programs.** Calling up a program from memory and reading "Dreamy Vocal" is much more descriptive than a number.

- **1 year warranty.** Why settle for 90 days if a company will support its products for a year or longer? Also, *make sure you send in the warranty card.* Sometimes companies offer software updates that improve a unit's performance, but they can't inform you of this if you haven't sent in your warranty.

Wall Wart Tamer

Wall transformers for gear can cut down on hum and simplify UL approval, but because of their size, they generally use up more than one receptacle on a barrier strip. Also, leaving them plugged in when no current is being drawn can shorten the transformer's life.

The wall wart tamer consists of a one- to two-foot length of two-conductor zip cord, and three parts that can attach to zip cord without any soldering: AC receptacle (GE Quick Clamp Connector, part #GE1710-21D, or equivalent), on-off switch (ACE Hardware Quick Connect Cord Switch, part #ACE 31089, or equivalent), and plug (ACE Hardware Quick Connect Plug, part #ACE 31099 or equivalent). Clip the receptacle and plug to opposite ends of the zip cord, and the switch in between (Fig. 1-1). Plug the transformer into the receptacle, and the plug into your barrier strip. Use the switch to turn off the AC when the unit isn't in use.

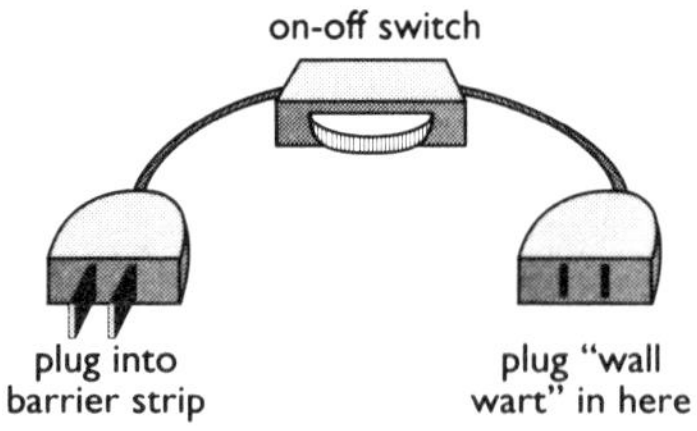

Fig. 1-1: The "wall wart tamer" reclaims outlets from barrier strips.

MIDI Features

- **Response to MIDI program changes.** This lets you use a sequencer or MIDI footswitch to call up particular programs. Because MIDI is a standardized protocol, you can use one manufacturer's footswitch with a different manufacturer's multieffects.

- **Response to MIDI continuous controllers.** You may not appreciate MIDI continuous controllers now, but someday you will (especially after you've read Chapter 5). They let you access individual effect parameters over MIDI, and control them with a specially-designed footpedal (or you can automate these changes if you drive the effects unit with a sequencer). Check that each patch can have its own continuous controller assignments, so that you can (for example) use a pedal to vary delay time in one patch, amount of treble boost in another, reverb depth in yet another, and so on. This lets you add a great deal of expression when playing live.

  Also check for smoothness of response—for example, if you vary the delay time does it change smoothly and predictably, or jump from one setting to another?

- **MIDI system exclusive (sys ex) dump options.** With multieffects, one or two hundred programs may seem like a lot of different sounds, but you can use them up faster than you might think—particularly if you develop custom libraries of sounds for multiple instruments. Being able to dump the memory contents over MIDI to a sys ex storage device (such as a computer running appropriate software, dedicated sys ex storage box, or a keyboard with sys ex storage capabilities) lets you recall a group of settings at any time. For more information on sys ex, see the end of Chapter 5.

- **MIDI thru jack.** If you have more than one MIDI-controlled effect, a MIDI thru jack simplifies distributing MIDI data to other units. The alternative is buying a separate MIDI thru box (and the fewer boxes you have to cart around, the better). Note that some units offer a "MIDI echo" or "MIDI merge" function that turns the MIDI out jack into a MIDI thru jack; this also does the job.

CHAPTER 2

Understanding and Programming Multieffects

Today's multieffects are woefully underutilized, partly because there is so much potential to explore, but also partly because these new devices intimidate many people. Familiar knobs and switches have been replaced by LCDs and little buttons, which demand a new way of looking at effects.

For example, few musicians seem to realize that contemporary multieffects allow for using a pedal, MIDI sequencer, and/or footswitch to vary virtually any aspect of a sound (overall level, number of echoes, reverb depth, distortion intensity, etc.) in real time. Many recording engineers don't exploit the fact that MIDI-controlled multieffects offer a type of highly sophisticated, yet inexpensive, automated mixdown. However, learning how to do these tricks can be confusing—unless you possess a few basic bits of knowledge that demystify what multieffects are all about.

In this chapter we'll delve first into how digital effects work, then describe a generic approach to programming multieffects that will help you get the sounds you want.

What Makes Muliteffects Tick

Analog and Digital Technology

Multieffects are an example of digital technology—the same technology that has brought us personal computers, programmable VCRs, and greeting cards that talk to you when you open them up. Until the 1980s, analog technology dominated the musical landscape. Digital technology performs many of the same functions as analog technology, but in general, offers more features at a lower cost.

Analog devices work directly on the audio signal itself. For example, if you want distortion, you take the audio signal to be distorted and overdrive an amplifier; for filtering, put the signal through a filtering circuit that has a limited frequency response. Unfortunately, since you're sending the signal itself though various pieces of circuitry, any noise or distortion generated by that circuitry becomes part of the signal, which degrades the quality. Digital electronics, which is based on computer technology, works quite differently.

Any audio signal, from a solo guitar to an orchestra, can be represented as a voltage that varies over time (Fig. 2-1). This is called a *waveform* because sound moving through the air resembles a wave, like ripples in a pond. Applying this waveform to a loudspeaker causes it to vibrate and move air; this creates changes in air pressure that we hear as sound (for more on the subject of waves and frequencies, see the section on Equalizers in Chapter 3).

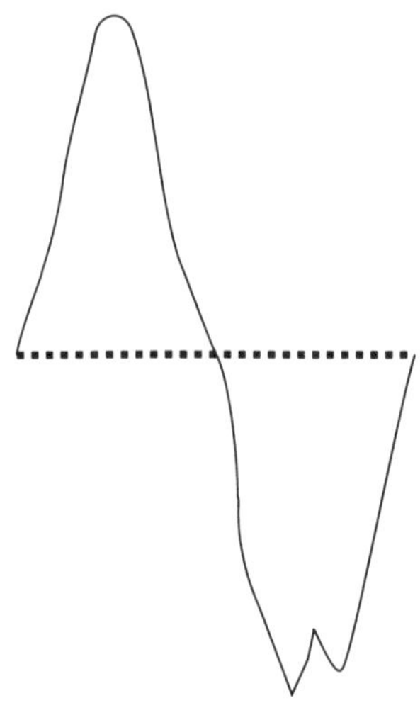

Fig. 2-1: A typical audio waveform (an orchestral violin section, in case you wondered).

If we measure the waveform's instantaneous voltage level at a very fast rate (say, 50,000 times per second), we'll generate a stream of numbers—*samples* of the waveform—that describes how the signal's voltage varies over time (Fig. 2-2).

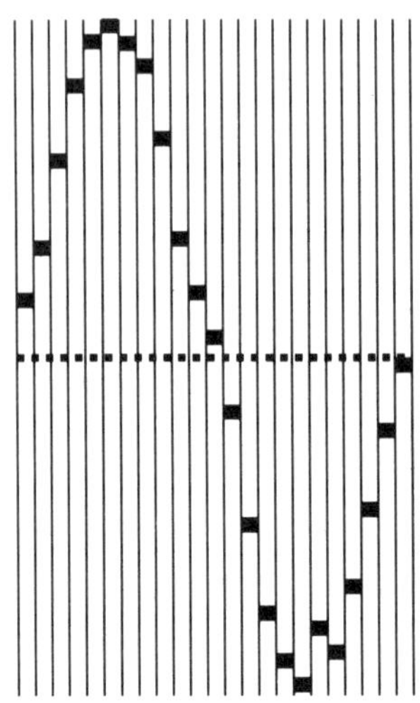

Fig. 2-2: The continuously varying analog waveform has been converted into a series of samples, which are then assigned numerical values.

Measuring each sample's voltage produces a series of numbers that defines the waveform numerically. The greater the number of samples that occur in a given period of time (called the *sampling rate)*, the greater the accuracy of the digitized signal. Higher quality multieffects sample at 44.1 kHz—the same sampling rate as a compact disc—or 48 kHz, which is the default sampling rate for digital audio tape (DAT) recorders. Some devices save money by sampling at 30 kHz; this still produces acceptable results.

So, we've successfully converted an analog (continuously varying) signal into a digital (series of numbers) signal, a process called *analog to digital conversion* or *digitization.* The same process can also work in reverse; *digital to analog conversion* transforms a series of numbers (which we can't hear directly) back into a varying voltage level so we can hear them.

Computers love to play with numbers, so after converting a signal into numbers it's time for computerized fun and games. Increasing the numbers' values provides amplification, decreasing the values causes attenuation, storing a series of numbers in computer memory then spitting them out periodically produces echoes, and so on. Because the computer is working on a *representation* of the signal, not the signal itself, the signal stays much cleaner than if it was going through an equivalent amount of analog circuitry.

The Analog Vs. Digital Controversy

As soon as digital technology appeared, there were boosters who felt it was the answer to everything, and naysayers who said it sounded horrible and would probably lead to the decline of civilization as we know it. As usual with extreme positions, the truth lies somewhere in between.

Although generalizations are always risky, there are differences between the two technologies that can affect the sound. Some of these are:

Distortion. When overloaded, analog units tend to distort in a more subjectively pleasing fashion than digital gear. Overloading analog circuitry yields an "overdriven" sound whose distortion products relate harmonically to the input signal, whereas overloading a digital delay produces distortion that is unlike analog distortion—it's more of a "splattering" kind of effect that bears little relationship to the signal feeding it.

Another difference is that analog units distort in a progressive manner. As a signal increases past the point where distortion begins, the distortion increases in a fairly predictable way. With digital technology, the signal will be virtually undistorted up to a certain point, then abruptly start to splatter. Turning the signal up past this point simply gives stranger, more dissonant splatterings. (Not that these don't have their place....)

One interesting aspect of some digital circuitry is that distortion actually *increases* at extremely low levels, unlike analog devices where distortion is lowest at low signal levels. This is due to a phenomenon known as "quantization error" (which you really don't need to know about if all you want to do is make or record music).

Fidelity. A digital device's audio quality depends directly on how well the unit can encode the analog signal at its input into digital information. Switching over to computer language, the encoder will be specified as having a certain number of "bits." While getting into a detailed discussion of bits and resolution is beyond the scope of this book, suffice it to say that the more bits, the better the quality. High-end studio gear will have 16 to 20 bit encoders, while lower-cost devices will have 8 to 14 bit encoders.

Noise. Analog and digital units have different "flavors" of noise. While this is pretty tough to describe in print, analog noise tends to be grainy, like a photograph shot with grainy film. Digital noise, on the other hand, is a more spiky kind of noise that tends to sound sharper and less smooth.

Understanding Programmability

The concept of programmability was introduced to musicians when synthesizer players became fed up with trying to change sounds rapidly on stage. Early, non-programmable synthesizers had so many knobs and switches they looked somewhat like a jetliner cockpit, and trying to call up a new sound in time for the next song drove many a player nuts. Programmable synthesizers let you edit a particular sound, then press a button to store the control setting information as a program in memory. Reselecting that program at a later time produces the same sound as when you laboriously adjusted the parameter values in the first place.

As signal processors became more complex, they also became more difficult to adjust in "real time." Once again, programmability provided a solution. Now we have multieffects with literally hundreds of parameters, but you can recall all their settings at the touch of a button.

However, programmability requires some adjustments in your thinking. Musicians and engineers are used to immediate gratification—bend a string, move a fader, or flick a switch, and the results are immediately apparent. And even with non-programmable rack-mount effects it doesn't take too terribly long to, say, turn up the feedback for a more intense sound. The downside of programmable effects is that they trade off the convenience of instant recall for the inconvenience of time-consuming programming to get the effect you want.

What makes programmability possible is the computer chip at the heart of every digital multieffects. Each effect or combination of effects is the result of a computer program that tells the computer how to create chorus, reverb, distortion, and/or other effects. This program was written by a real live human being, but *you* edit it. When you're getting the sound you want out of a multieffects devices, you're actually entering data into a computer program to change what the program will do. (Actually, to say that multieffects have a computer inside is a bit of an oversimplification. Most devices have a special purpose computer called a Digital Signal Processor [DSP] that is optimized for processing audio signals. DSP chips are very sophisticated "number crunchers" that can do the complex, real time calculations needed for audio work.)

For example, echo consists of delaying a signal, feeding some of the delayed output back to the input to create additional echoes, and mixing some echoed sound in with the straight sound. So, a multieffects' echo program would tell the computer to "delay a signal for X milliseconds, feed back Y% of the delayed signal back to the input, and mix in Z% of echoed signal." The nature of the echo sound will change according to what data we put in for X, Y, and Z. A larger value of X means more milliseconds, thus a longer delay. If we feed back a small amount of signal (small Y value), we'll hear only a few echoes; larger Y values feed back more of the signal, creating more echoes that take a longer time to fade out. What we enter for Z determines the straight/processed mix.

Upon entering the data necessary to get the sound you want, you've created a variation which is also called a program (yes, jargon can be confusing). Generally, when we talk about a unit that "stores 100 programs," we don't mean the computer program the design engineer wrote, but rather the edited versions you've created. From now on when we say program, we'll mean your variations. These are also called *patches* or *presets,* and just to get you used to the real world, we may use these terms as well.

About Parameters

Each adjustable element of an effect, whether analog or digital, is called a *parameter.* For example, the variable parameters on an analog delay line might include initial delay time, feedback, modulation depth, etc. Before digital electronics took over the world, an effects box had one control (switch or knob) per parameter, so changing parameter values was a relatively easy process. But knob-based effects also had problems: changing a sound (which required a lot of knob-twisting) took time, and if you came up with a great sound, trying to get it back later could be difficult. Also, knobs and switches have always been some of the most expensive components in effects.

Digital electronics largely eliminates knobs. Remember our example above where we described an echo sound with X, Y, and Z parameters? With a digital effects unit, each parameter would be given a unique name or number (so you could identify it for editing), and be *quantized* into a series of discrete steps (Fig. 2-3). For example, delay time, instead of being continuously variable and selected by a knob, might be quantized into 1 millisecond steps and selected by keying in a three-digit number *(e.g.,* 000 to 999 milliseconds) with a keypad.

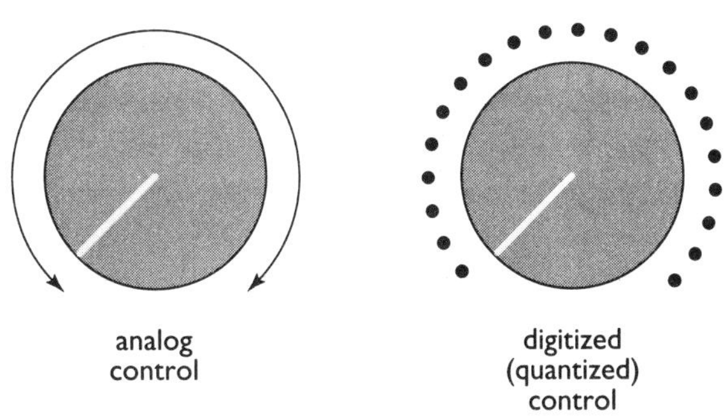

Fig. 2-3: Analog controls are continuously variable. A digitally-controlled parameter is like a knob that has been divided (quantized) into a series of discrete steps.

The reason for quantizing parameter values is because once these are identified numerically, the same computer doing all the other tasks we described earlier can turn its attention to storing these values in its memory. This lets us call up a particular program at any time. For example, suppose we told the computer that our echo program's X parameter was 210 milliseconds, Y parameter 40% feedback, and Z parameter 35% echoed signal. The computer can remember this group of numbers as a program; once you give the program itself a number, like 26, the computer will file all this information in its little brain under "26" so that next time you ask it for program 26, all parameters will be reset exactly as you specified.

Sounds good so far... now it's time to figure out how to *access* all these parameters.

Multieffects Programming

Accessing Parameters

As mentioned earlier, most computerized musical devices don't have knobs that you can twist to change sounds; instead, you need to find individual parameters and alter their values, usually by a process of button-pushing.

Fortunately, there are only so many ways to accomplish a given task. If you're creating a sound from scratch or editing an existing sound, you're almost always going to use the same basic procedure for any device:

1. Specify the program (patch) to be edited. This reserves a memory location that temporarily holds the parameter edits.

2. Select the program's structure, which is called an *algorithm*. The algorithm will determine the sound's overall character. You may have a choice of several fixed algorithms *(e.g.,* compressor » distortion » chorus » EQ » reverb) or you may be able to choose the order and type of effects. Fig. 2-4 shows a couple different algorithms that define the effect's structure.

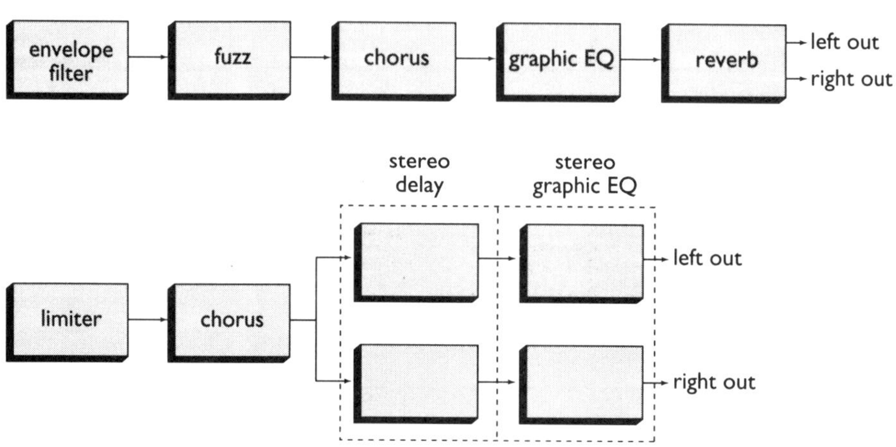

Fig. 2-4: Two different algorithms. Each one creates a different type of effect, and has variable parameters so you can alter the sound.

3. Specify a parameter within the algorithm that you want to change (echo time, amount of distortion, noise gate threshold, etc.).

4. Enter a new parameter value and listen to what effect this has on the sound.

5. Repeat steps 3 and 4 until all the parameters have been adjusted to give the type of sound you want.

The most common data entry tools are a calculator-style keypad for entering numbers, and/or scrolling or "arrow" keys to help locate the different parameters (we'll see how these work in a little bit).

Selecting Different Programs

A very basic function on all units is calling up different programs. When you turn on a digital multieffects, odds are you'll be greeted with either the last program you selected or a default program *(e.g.,* program #01). Depending on the unit, to select a new program you might punch in a certain program number with the keypad, or scroll through the different programs with the arrow keys.

You can think of the programs as forming a list, with a window that scrolls over the list (Fig. 2-5). The up and down arrow keys move the window over the list to select a particular program. For example, if you're on program 14, pressing the up arrow key selects program 15; pressing the down arrow calls up program 13. Some units may use a knob to select programs instead. Then again, some devices may arrange their programs "horizontally" instead of in a vertical list, and use right/left arrow buttons to move from one program to another. In either case, the basic principle remains the same.

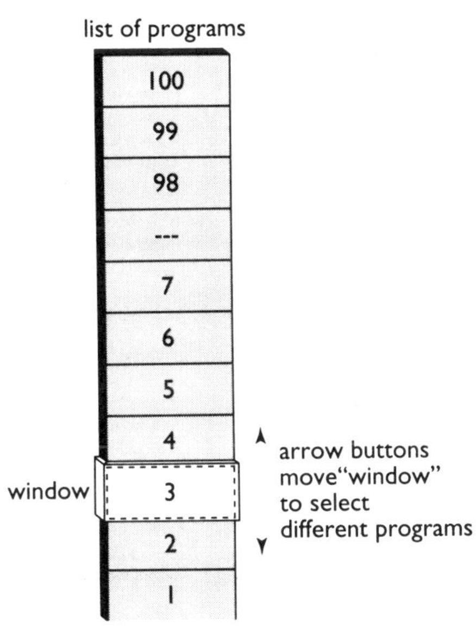

Fig. 2-5: Using up/down buttons to select a program.

Parameter Selection and Editing

Now that we have a program, it's time to select and edit parameters. Each unit has a slightly different way of doing things, but here's a typical real-world example based on a generic multieffects.

Suppose a multieffects has two displays (left display for program number, right for other parameter values), two sets of up/down buttons, and one set of left/right buttons. You would begin by selecting a program, as shown on the left display, with the first set of up/down buttons (Fig. 2-6). The up button selects the next higher-numbered program, and the down button, the next lower-numbered program.

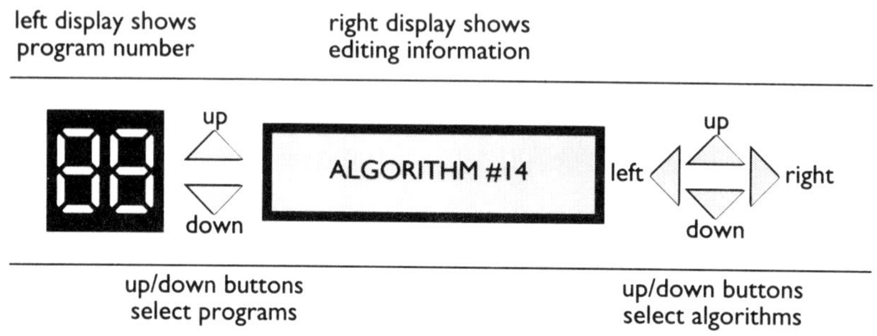

Fig. 2-6: Program and algorithm selection with a generic multieffects.

Upon calling up a program, the right display might then show the name, number, or even a block diagram of the algorithm used in the selected program (remember, each algorithm represents a particular combination of different effects). If you wanted to choose a different algorithm, you could do so with the second set of up/down buttons.

Each algorithm has an associated "list" of parameters. Since we used the right-hand set of up/down buttons to select an algorithm, it follows that we'll use the left/right buttons for the next step—parameter selection (Fig. 2-7).

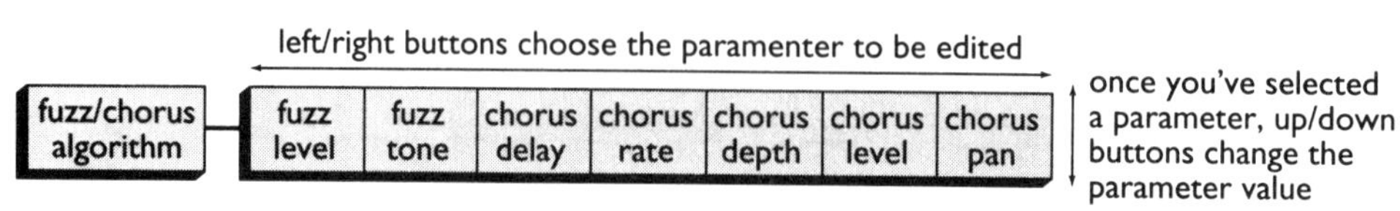

Fig. 2-7: Each algorithm will have several parameters whose values you can change.

As you press the left or right button, the display identifies the selected parameter. To change the parameter value, use the right hand set of up/down buttons; the display will show the parameter's value. After editing the value, press the left or right button again to select the next parameter on the list (Fig. 2-8).

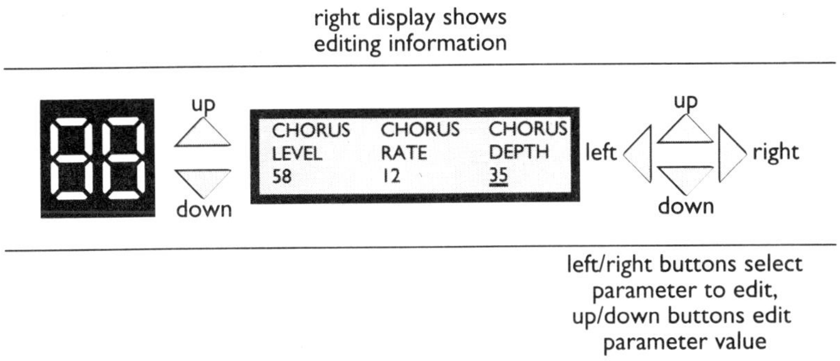

Fig. 2-8: In this example, the display shows several parameters (other multieffects may show only one parameter per screen). You select a parameter for editing with the left/right buttons; the one being edited has an underline (cursor). In this case, it's chorus depth, which has a value of 35. The up/down buttons change the parameter value.

This process illustrates two important points:

- There is a definite order for parameter editing. You must first choose the program, and if applicable, the desired algorithm before you can choose a parameter whose value you want to change.

- Note how the display anticipates your needs. If you press one of the right hand up/down buttons with an algorithm number showing, the display knows that you want to edit algorithms. If you press one of the left/right buttons with an algorithm number showing, the display knows that you want to edit the parameters within that algorithm. If you have a parameter selected and you press an up/down button, the display knows you want to edit values.

This demonstrates the good news/bad news of digital effects: if you know what you're doing, editing flows in a logical fashion. If you don't know what you're doing, and you press the wrong button at the wrong time, you may get lost in the program and not be sure what you're adjusting (or how to get back to a familiar reference point).

Different units use variations on a theme. Some boxes have dedicated buttons for turning individual effects in an algorithm on and off ("effects select" buttons). If you wanted to edit one of these effects, you might press an "edit" button to select a parameter editing mode, which would then change the effects select buttons into effect edit buttons. For example, pressing the compressor on/off button while in edit mode would select the compressor for editing. Each successive press of the compressor button would access another compressor parameter, and up/down arrow buttons would set the value.

Although the specifics are different from the example given above, you still:

- Select a program
- Select an algorithm or effect
- Specify a parameter
- Change the parameter's value.

No multieffects unit (or synthesizer) strays too far from this basic concept. Once you figure out how your multieffects performs these steps, you're on your way to being a programming expert.

Making Your Own Algorithms

Although many multieffects string certain effects together in algorithms, some devices let you create your own algorithms out of an available palette of effects. This gives considerably more flexibility, but requires more of an effort to learn. For more information how to get the most out of this type of effects architecture, see Chapter 4.

Shortcuts

Since button-pushing is tedious, manufacturers often include shortcuts. For example, scrolling through 99 programs with up/down arrows can take some time. So, one unit might increase the scrolling rate the longer you hold the button down, while another might double the scrolling rate if you press the unused arrow button while holding down the desired arrow button. And since even little buttons cost money, a manufacturer might use a "shift" button (like the shift key on a computer keyboard or typewriter) that changes the function of a set of buttons so that five buttons and a shift button can do the work of ten buttons. Any shortcuts should be documented in the manual.

Parameter-controlled effects may be confusing at first, but don't give up. You have a lot more power at your fingertips, and greater repeatability. Sure, it takes more time to program or tweak a sound initially, but once you find a great sound and store it in memory, you won't have to find it again.

CHAPTER 3

Multieffects Parameters Explained

To program a multieffects and get your own unique sound, you need to understand how each of the dozens of available parameters influences what you hear. This isn't as difficult as it may seem, since you can always just select a parameter, change it, and listen to what happens; after a while, you'll figure out what each parameter does.

However, trial-and-error can be a time-consuming process. To make life easier, this chapter explains the functions of the most common parameters you'll find in a typical multieffects unit. Of course, your particular unit may not include all these parameters, and it may offer parameters that aren't covered here—but that's what the owner's manual is for. After you become familiar with basic parameter functions, read the manual and find out more about what your unit can do.

The parameters are arranged by effect (distortion, equalization, compression, etc.). There's a brief description of what the effect does, followed by a listing of typical effect parameters, and where appropriate, tips on getting the most out of the particular effect. First, though, let's look at how to get the right amount of signal into the multieffects as well as how to bypass the effects altogether.

Setting the Input Level

Multieffects generate a certain amount of noise ("hiss"). Since the signal-to-noise ratio represents the amount of signal going through a unit compared to the noise generated by the unit, feeding in the highest signal level possible (short of distortion) gives the highest signal-to-noise ratio and the most noise-free operation. (For more about signal-to-noise ratio, see the sidebar "About Decibels" in the section on Compressor/Limiters.)

Virtually all multieffects include an *input level* control (also called input, input gain, sensitivity, etc.) to vary the amount of signal going into the unit. For weaker signals, turn up the input control to send more signal through the multieffects. With stronger signals, turn down the input control to prevent distortion. (Since controls are costly, some devices might simply have a parameter, accessed with the same buttons that select other parameters, to set the input sensitivity.)

Inexpensive multieffects typically include a single LED to indicate an overload condition. If this LED lights up frequently, decrease the input level somewhat. If the LED lights very rarely or not at all, then increase the input level to send the maximum possible signal into the device. More costly units often use a bicolor LED or multiple LEDs (like the LED VU meters on tape recorders) to indicate not just the overload point, but also, how close the signal is to approaching that point.

Note that some LED meters are switchable between monitoring the input signal and the digital signal processing section's output. This is because the *input* level might fall within an acceptable range, yet the *processing* could create an overload situation (such as what would happen if you added a huge amount of equalization boost). If you have a choice, it's generally best to monitor the DSP output.

Some rack-mount multieffects are not really designed to handle relatively weak input signals (such as microphones and stock, non-active guitar pickups). Attempting to use one of these signal sources will not damage the multieffects or instrument but might lead to a duller sound and/or increased noise. Should you encounter these problems, insert a preamp, buffer, or other active electronic circuit with a high input impedance (greater than 100kΩ or so; check the spec sheet) between the instrument output and multieffects input. This isolates the instrument from the multieffects and gives a cleaner sound.

The *bypass switch* chooses between the straight, unprocessed sound and whatever effect you have set up on the multieffects. Many units include a jack on the back for plugging in a remote bypass footswitch; this feature is useful when both hands are occupied. It is also often possible to use a MIDI command to enter bypass mode.

Compressor/Limiter

A compressor evens out variations in dynamic range by amplifying soft signals to make them louder, and attenuating loud signals to make them softer. The net result is much less level difference between soft and loud signals, which evens out level variations and increases the sustain of percussive instruments.

One way to achieve a crude simulation of compression would be to carefully operate a manual volume control at the output of a standard preamp: as the level increased, you would turn down the volume control to compensate, and as the level decreased, you would turn it back up. Although this may not sound too sophisticated, many recording engineers use a variation of this technique (called *gain-riding)* to keep sudden bursts of volume from saturating the tape, or quiet passages from getting buried in noise. However, this clearly isn't a very precise way of doing things, nor can gain-riding respond to rapid changes in signal level (unless you have *very* fast hands!). Electronic compression is much faster and offers additional options.

Fig. 3-1 graphically shows the amplitude of a decaying guitar string; Fig. 3-2 shows the same signal after compression.

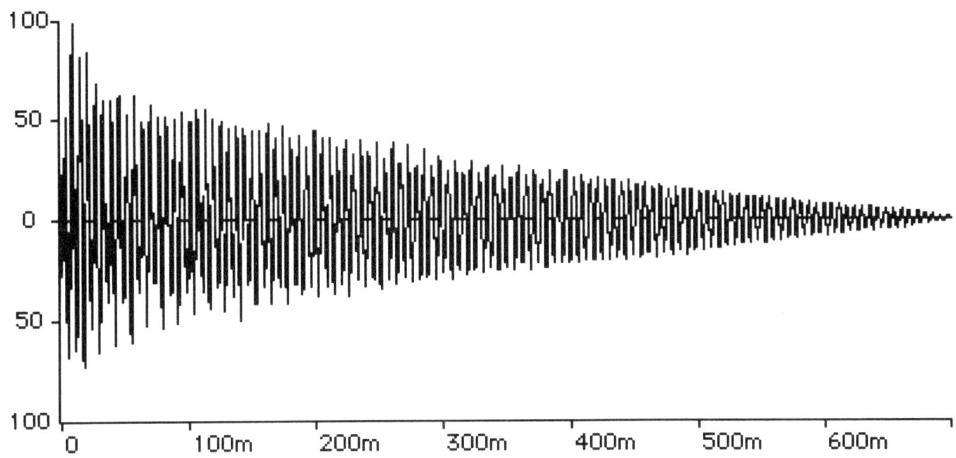

Fig. 3-1: A decaying guitar string. Note how the amplitude starts off strong, then decays to nothing.

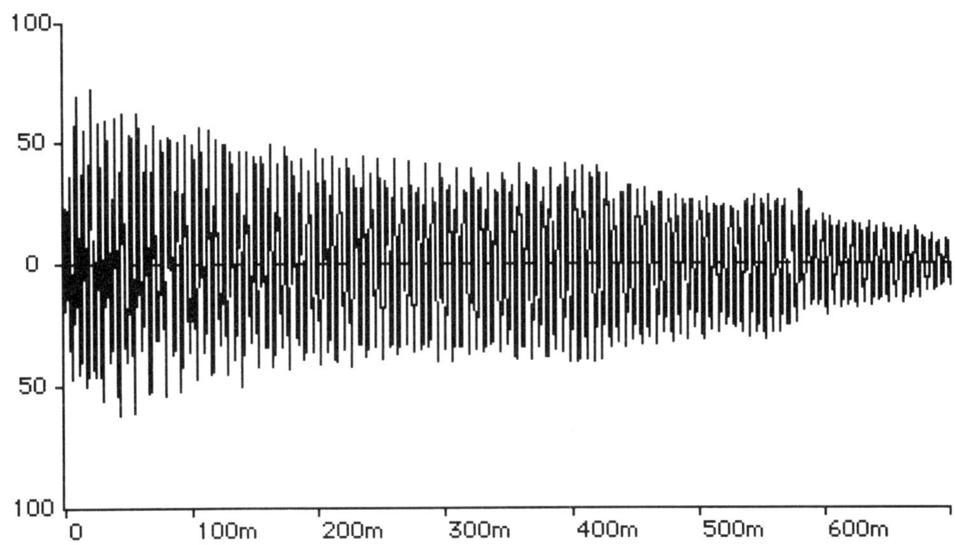

Fig. 3-2: After compression, the peak is slightly attenuated and the decay is greatly amplified.

About Decibels

The decibel (dB) is the unit of measurement of the ratio of two audio signals. For example, a device's signal-to-noise ratio (the amount of signal compared to the amount of noise) is typically expressed in dB, as is the amount of gain through a system *(i.e.,* the ratio between the system's output compared to the signal at the input). The decibel can also quantify level changes at a device's input or output; for example, you could increase the input level by a certain number of dB, or apply gain to a preamp to increase the output level by a certain number of dB. Let's examine a device's signal-to-noise ratio as a way to explain the concept.

All electronic devices generate noise, which shows up at the output. When you compare a device's maximum available output signal to the amount of noise present at the output, you define the signal-to-noise ratio, or S/N ratio for short. For example, if there's a 1 volt audio signal at the output of a multieffects, and the residual noise output is 1 millivolt (1/1000th of a volt), then your S/N ratio is 1000:1.

The decibel has a logarithmic curve to more closely follow the way the ear hears, which is very sensitive at low levels and less sensitive at high levels. *Doubling* the ratio of one voltage to another *adds* 6 dB (likewise, halving a voltage ratio subtracts 6 dB). For example, a ratio of 1000:1 corresponds to 60 dB; a 2000:1 ratio is 66 dB, a 4000:1 ratio is 72 dB, an 8000:1 ratio is 78 dB, and so on.

The larger the S/N ratio, the lower the noise. Many times the S/N ratio will be given as a negative number (such as -70 dB) since technically speaking, the noise is lower (hence the negative number) when referenced to the audio signal. So, you can either think of the audio signal as being +70 dB above the noise, or the noise as being -70 dB below the audio signal.

The decibel is also useful when applied to gain. If a preamp provides 6 dB of gain, that means

A *limiter* is similar to a compressor since it restricts dynamic range, but it affects only the signal's peaks. A limiter will not let a signal exceed the level specified by a *threshold* parameter, but leaves lower-level signals untouched (Fig. 3-3).

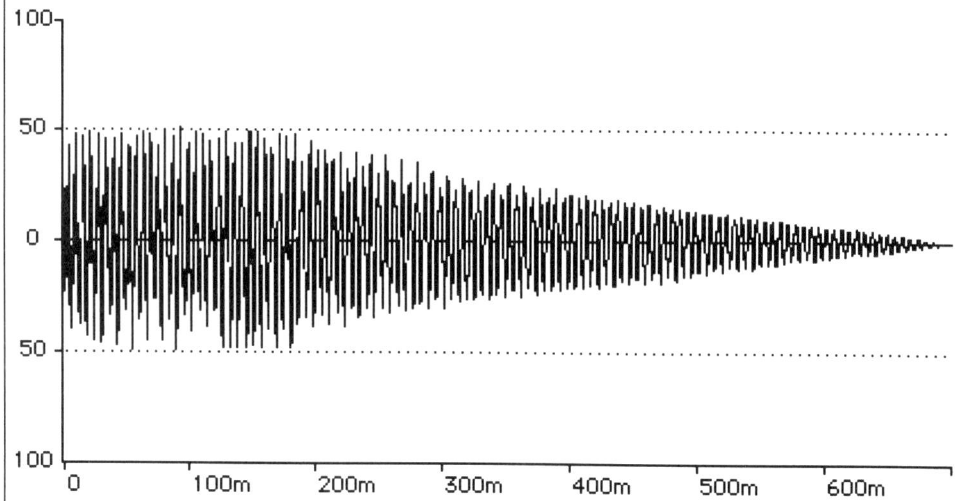

Fig. 3-3: After limiting, anything above the threshold (in this case, it has an arbitrary value of 50) gets clamped to that threshold. Lower level signals are unaffected.

Modern compressors are often a hybrid between a "classic" compressor and limiter, and are referred to as compressor/limiters (C/L for short). With C/Ls, you set a threshold above which compression occurs. Once a signal exceeds this threshold, increasing the input signal by a certain amount gives a lesser increase in output. For example, increasing the signal level by 6 dB (decibels; see the sidebar) may yield an output increase of only 3 dB. This would be considered a *compression ratio* of 2:1 *(i.e.,* for every 2 dB of input level increase, there's only 1 dB of output level increase). If increasing the signal by 6 dB gave an output increase of 2 dB, that would be a 3:1 compression ratio.

Compressor/Limiter Parameters

A typical C/L has several parameters which often interact *(i.e.,* after adjusting one parameter you may need to go back and tweak a different parameter).

Threshold sets the level above which signals will be compressed or limited. The lower the threshold, the more the signal will be compressed or limited, and the greater the sustain. If the signal drops below the threshold, then the C/L "goes back to sleep" and leaves the signal alone until it exceeds the threshold again. (Typical threshold control settings for guitar are -3 to -25 dB.)

Ratio selects how the output signal changes in relation to the input signal once the input signal exceeds the threshold (Fig. 3-4). For example, as mentioned earlier a ratio of 2:1 means that an input volume change of 2 dB gives only 1 dB of output change. The higher the ratio, the greater the amount of compression, and the more "squeezed" the sound. Extremely high ratios clamp the output level to the threshold. This puts a "ceiling" on the signal, regardless of input level changes, and causes the C/L to act like a limiter.

the output voltage is twice as much as the input voltage.

The subject of decibels is quite complex, since there are different types of decibels used for different types of measurements. However, just remember that it expresses the ratio of one audio signal compared to another, and you'll have the basic idea.

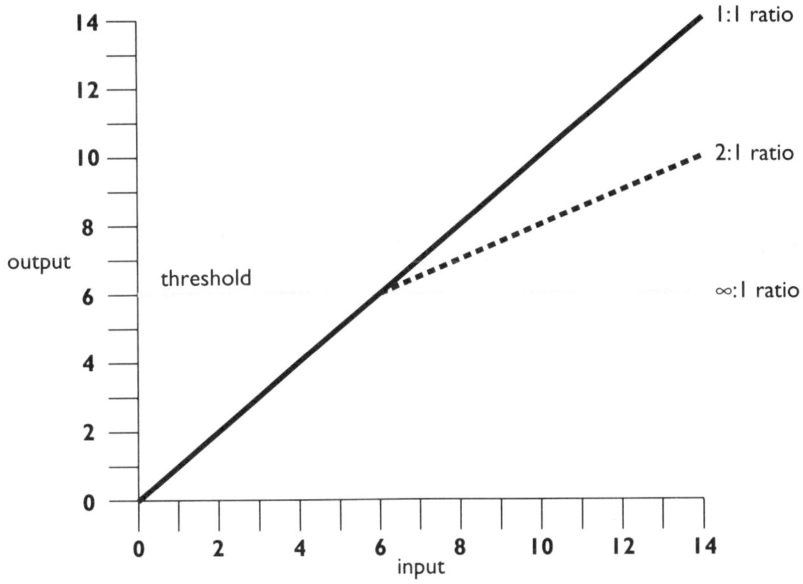

Fig. 3-4: With compression/limiting, the threshold sets the point above which compression occurs.

Output adds gain to offset the lower level caused by squeezing the dynamic range.

Attack sets the time it takes for the C/L to react to input level changes. A longer attack time "lets through" more of a signal's original dynamics before the compression kicks in. (For example, if you wanted to retain the initial "thwack" of a kick drum or the pick noise in a compressed guitar sound, you'd add a bit of attack time.) A typical setting is 0-30 ms.

Release determines how long it takes for the C/L to return to its normal state after going under the threshold. With short release times, the C/L tracks even very slight level changes. This can produce a "choppy" sound; turn up the release time to smooth things out. 75-250 ms seems to work well with guitar and vocals.

Response determines how the C/L reacts to input signal level changes. With *peak* response, the C/L monitors the instantaneous signal level. Therefore, as soon as it gets hit with a peak—like a pick transient—the compression/limiting action is immediate (subject to the attack time control setting). *Average* response monitors the average signal level over a short time period, and uses this signal to control the gain. Average response is used more when compressing program material, such as a song you're mixing.

Compressor/Limiter Tips

- **Minimizing noise.** Do not overcompress, as this makes for a thin, unnatural sound. Use the bypass switch to compare the compressed and non-compressed sounds; you may find that even a little bit of compression gives the desired effect. Generally, a 2:1 ratio works well for general applications. Higher ratios *(e.g.,* 8:1) are best when you really want to sustain a sound.

- **Optimum signal chain placement.** With multieffects that let you move effects around in the signal chain, place the compressor toward the beginning so that it doesn't bring up the noise from previous stages.

- **Guitar sustain.** When increasing guitar sustain, remember that compressor/limiters are not miracle workers. They cannot make your guitar's strings vibrate any longer, but can only increase the apparent sustain. Don't think that a compressor can compensate for dead strings or for guitars with poor sustain characteristics.

- **Smoother distortion sound.** Add compression before distortion for a smoother sound with more sustain.

- **"Mystery" compression increases.** If it seems like there's been a sudden increase in compression but you didn't increase the compression amount, then the input signal going to the compressor may have increased.

- **Compressing lead guitar but not compressing rhythm.** You can take advantage of the fact that increasing the input level increases compression when you want a sustaining lead but an uncompressed rhythm sound. Turn up your guitar's volume control for leads, and turn back down for rhythm. As the guitar signal drops below the threshold, the compression will either go away, or have much less of an effect.

- **That giant sucking sound.** Some music from the 60s, particularly psychedelic-era Beatles recordings, featured a drum sound that sounded like it was "sucking" and inhaling. To create this effect, apply lots of compression to the drums with an extremely short release time.

Distortion

Distortion mimics the way an amplifier behaves when overloaded, so it's one of the most popular effects in a guitar player's repertoire. However, distortion can do much more than grunge out a guitar—some top artists also use distortion on sounds such as drums, synthesizers, and even vocals.

Not all types of distortion (tube, transistor, digital, etc.) sound the same, and there's no real consensus about what makes for "ideal" distortion. As a result, one of the most variable elements of a multieffects is the way it distorts ("clips") a signal. Some devices include an actual tube stage or other analog distortion circuit that can be modified under computer control. Others use DSP to emulate particular types of distortion sounds.

Hard vs. Soft Clipping

With hard clipping, the output signal remains undistorted up to a certain point, then becomes extremely distorted as the input increases past that point. Standard transistors tend to distort in this manner.

Most musicians tend to prefer "soft" clipping, as exemplified by tube and analog tape distortion. With soft clipping, the output signal becomes progressively more distorted as the input signal level increases (Fig. 3-5). This type of distortion is more "natural" in the sense that it creates a feeling of dynamics—bashing on a guitar string harder will give more distortion, and the distortion's timbre will change as the string decays. With hard clipping, there isn't too much variation in the sound until the output signal passes below the clipping point.

To test the clipping characteristics, select a program that uses distortion and turn the input level to the multieffects down until the signal is on the verge of distortion. If there's a smooth transition to greater distortion as you increase the input, the device tends toward soft clipping. If the signal becomes "spiky" as it passes the point where distortion begins, there's more of a hard clipping effect.

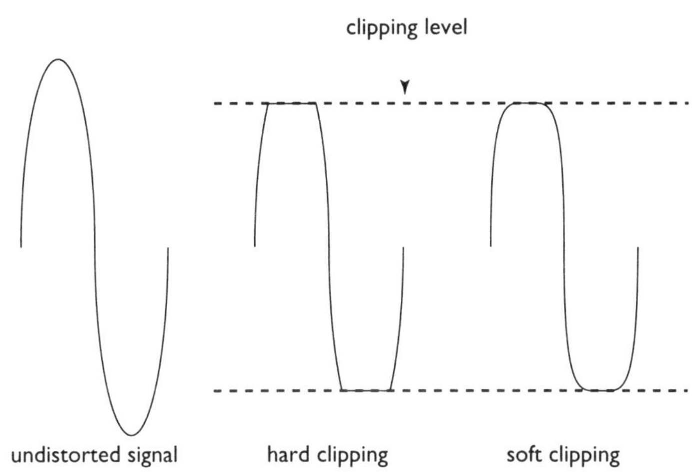

Fig. 3-5: An undistorted signal compared to hard- and soft-clipped versions.

Typical Distortion Parameters

Sensitivity, drive, or *input* determines the amount of signal level needed for the onset of distortion. At maximum sensitivity, the distortion effect will be more pronounced, while at minimum sensitivity the effect will be less intense.

Clipping level works in conjunction with the sensitivity control. The lower the clipping level, the lower the signal level needed to create distortion. Higher clipping levels require higher signal levels before distortion occurs.

Output. Since distortion often adds a great deal of amplification, the output parameter lets you trim the effect's output level. Although you can set this for unity gain *(i.e.,* the distorted sound is subjectively as loud as the straight sound), musicians commonly set the distorted level somewhat higher than the straight signal for a boosted, "in your face" sound.

Tone controls. For added flexibility, some distortion effects include built-in tone controls. Distortion adds harmonics to the signal, which increases the high frequency content; pulling back on the high frequencies reduces shrillness, while boosting the bass gives more depth and power. For more information on tone control parameters, see the following section on equalizers.

Noise gate. Because distortion adds so much gain, there's often increased noise and hiss. A noise gate helps remove this by shutting off signals below a certain level. (Even if the multieffects has a separate noise gate module, the distortion effect may have its own dedicated noise gate.) For more information, see the section on noise gates.

Distortion Tips

Since distortion is most often used by guitarists, here are some suggestions on how to set guitar controls for particular distortion qualities.

- **Smooth sound.** Use the bass pickup with the tone control set for minimum treble.

- **Raunchier sound.** Use the treble pickup.

- **Using optional guitar switches.** If your guitar has a series/parallel pickup switch, the series position will give the thickest fuzz sound. If your guitar has an in phase/out of phase switch, the out of phase position will give the thinnest sound.

- **Brighter sound.** Some guitars (particularly Fender solid body types) have a feature where turning down the volume control attenuates the low frequencies at a faster rate than the high frequencies. Thus, if you turn the control down about 3/4 of the way, the upper strings will distort more than the lower ones. This gives a bright, lively type of distortion.

- **Balancing the sound.** With guitar, pickup height adjustments are crucial to getting a consistent distortion sound. If the sound is too boomy, angle the pickups slightly so that the bottom three strings are further away from the pickup than the top three strings. Also, note that newer strings will sustain longer than older strings.

With signal sources such as electronic drums, it's best to set up a submix of the sounds you want distorted, since distortion has varying degrees of effectiveness with different drum sounds. For example, start with the toms and kick, then listen to what happens when you mix in snare, high hat, etc.

Distortion can also help give ultra-clean synths a bit more personality. Add a hint of distortion to organ patches with rotating speaker effects, or to that old DX7 keyboard that's sitting around and feeling neglected. You might be amazed at how a *little* distortion can make a synthesizer sound a lot more "rock and roll."

Equalizers

An equalizer emphasizes or de-emphasizes certain frequencies (see sidebar) to change a signal's timbre. For example, emphasizing the bass frequencies gives a muffled, rumbly sound. Boosting high frequencies creates a brighter, more shimmering sound; boosting midrange frequencies adds punch and definition. Cutting (also called *de-emphasizing)* certain frequencies can alter the timbre of an instrument just as much as boosting, if not more so. The amount of boosting or cutting will usually be given in decibels (dB).

Equalization is not always intended to create special effects. Few instruments have a flat frequency response (guitar pickups tend to produce more low frequency energy than high frequency energy, stringed instruments may have "dead spots" on the neck where the response is not as strong, acoustic instruments can have extreme response peaks, and so on), so one of the most common applications of frequency response altering effects is to *equalize* some of these frequency response differences and create a more uniform sound.

Equalizers are based on filters, which pass certain frequencies and reject others. The four most common filter types are:

- **Lowpass Filter.** This passes all frequencies below a certain frequency (called the *cutoff* or *rolloff frequency),* while rejecting frequencies above the cutoff frequency. In real world filters, this rejection is not total. Instead, past the cutoff frequency, the high frequency response rolls off gently (Fig. 3-6). The rate at which it rolls off is called the *slope.* Note that some digital filters can provide an extremely abrupt slope; these are called *brickwall filters* because anything higher than the cutoff frequency hits a "brick wall."

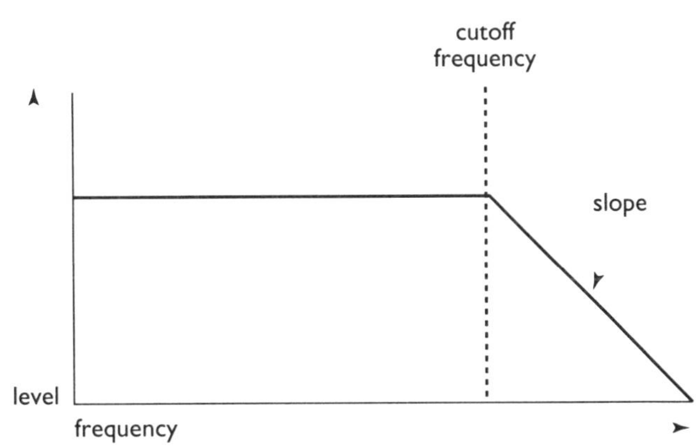

Fig. 3-6: Lowpass filter response.

Frequencies and Hertz

Sound moves through the air in waves, like ripples through water. When a sound generator such as a tuning fork vibrates, it sends out waves of air that hit our ears, and which we perceive as sound. Speakers are designed specifically to transform waves of electricity passing through the voice coil into moving air.

The number of waves that occur in a second is measured in Hertz (Hz for short), a unit of frequency measurement named after Heinrich Hertz (1857-1894), a German scientist who did extensive research into wave phenomena. This term replaces *cycles-per-second,* or *cps,* which was used prior to the 1960s.

Here are some examples: an A 440 tuning fork vibrates at 440 Hz; the AC power signal coming out of the wall has a frequency of 60 Hertz. A guitar's lowest note is approximately 90 Hz, and the highest, around 1,000 Hz. However, note that these figures represent only the signal's *fundamental* frequency, where most of the energy occurs. Additional, higher-frequency *harmonics,* which are related mathematically to the fundamental, create the distinctive timbres of various instruments. Bright, brassy sounds are rich in harmonics; dark, muffled sounds have fewer harmonics.

To express higher frequencies numerically, there's the kilohertz (kHz, which equals 1,000 Hz) and megahertz (MHz, or 1,000,000 Hz). The human ear responds to a range of frequencies from around 20 Hz to 20 kHz, assuming you're young and haven't damaged your hearing.

Equalizers, by boosting or cutting various frequencies, can alter an instrument's timbre subtly or radically. Equalization is one of the most important effects for creating specific sound qualities.

- **Highpass filter.** This is the inverse of a lowpass response: it passes frequencies above the cutoff frequency, while rejecting frequencies below the cutoff. Fig. 3-7 illustrates high pass filter response.

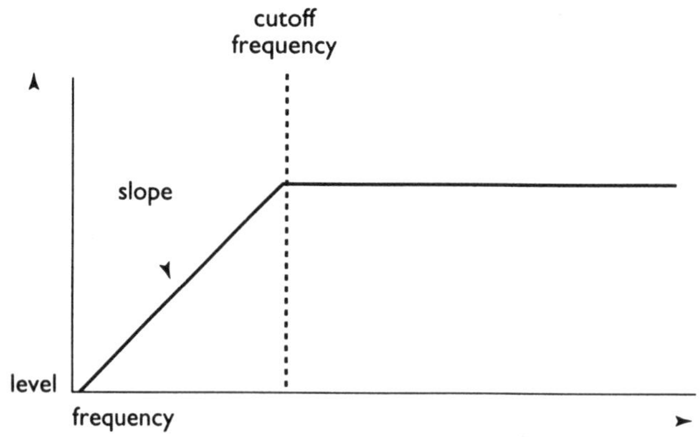

Fig. 3-7: Highpass filter response.

- **Bandpass filter.** This boosts only those frequencies around its *resonant frequency,* while rejecting higher and lower frequencies. Fig. 3-8A shows a bandpass filter with a gentle slope (less resonance), while Fig. 3-8B shows a bandpass filter with a steep slope (more resonance).

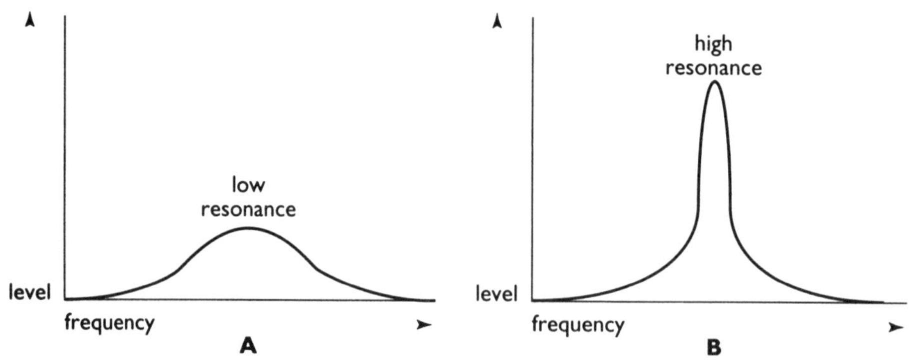

Fig. 3-8: Bandpass filter response, with two different resonance characteristics.

- **Notch Filter.** This response is the inverse of a bandpass: all frequencies around the *notch* frequency are rejected, while frequencies higher and lower than the notch frequency pass unimpeded to the filter output. Fig. 3-9A shows a notch filter with low resonance; Fig. 3-9B shows a notch filter with high resonance. The greater the resonance, the deeper the notch.

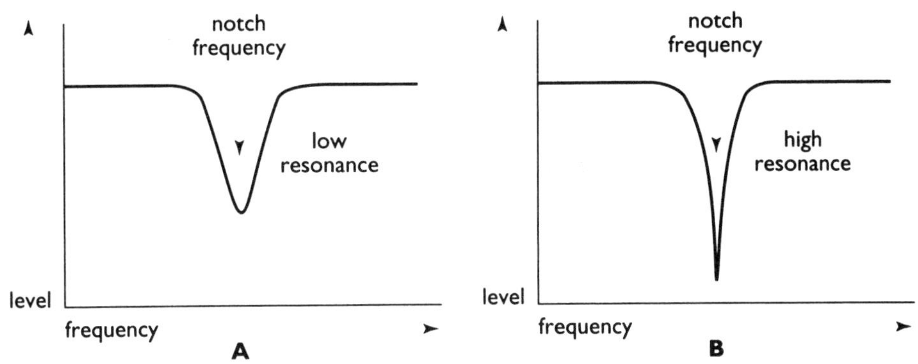

Fig. 3-9: Typical notch filter responses.

Lowpass and highpass filters often include a resonance parameter. Increasing the resonance creates a response peak at the cutoff frequency, thereby creating a "sharper" sound (Fig. 3-10). With extreme amounts of resonance, the filter can oscillate, which produces "whistling" effects.

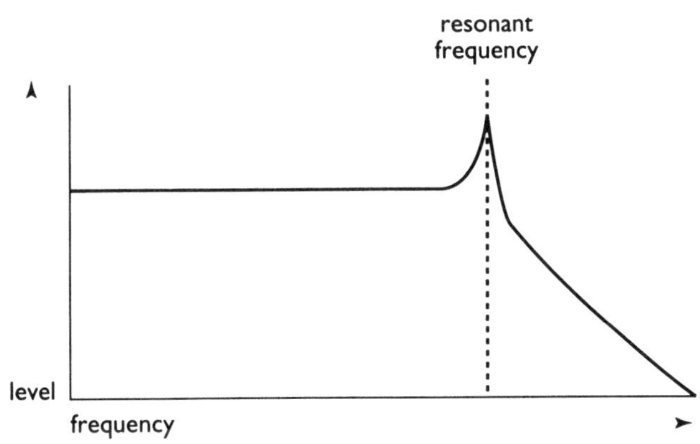

Fig. 3-10: A lowpass filter response with added resonance.

Types of Equalizers

The simplest equalizer is the standard tone control found on many electric guitars. The most common design is a simple lowpass filter that reduces the high frequencies, which gives a bassier sound. However, this is a fairly limited approach that doesn't come close to the flexibility of the equalizers found in multieffects.

The next step up on the equalizer food chain is the *shelving equalizer.* This is common in stereo rigs and smaller mixers; the most popular configuration includes two knobs, one to boost or cut bass, the other to boost or cut treble (Fig. 3-11). This doesn't allow for very precise sound shaping, but can correct broad deficiencies like lack of high frequency "sparkle" or excessive "boominess" in the bass end.

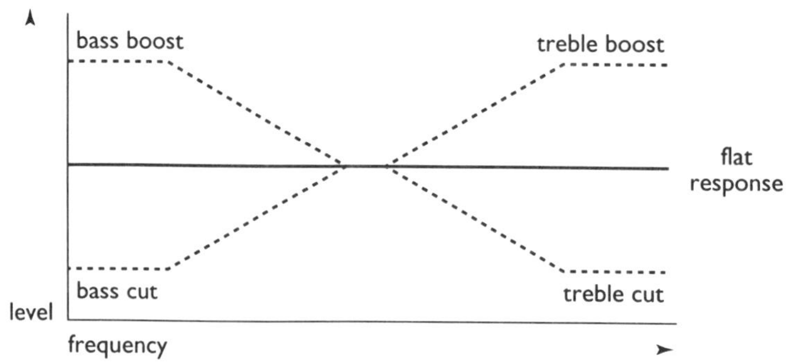

Fig. 3-11: Shelving equalizer response.

The *graphic equalizer* uses multiple bandpass filters to split the audio spectrum up into a number of bands, with an individual boost/cut control for each band. The term graphic equalizer refers to the fact that hardware graphic EQs use linear slide pots for the boost/cut controls, so looking at the position of the knobs gives a "graphic" indication of frequency response (Fig. 3-12). Some multieffects graphic equalizers show a "curve" on the LCD to maintain this tradition.

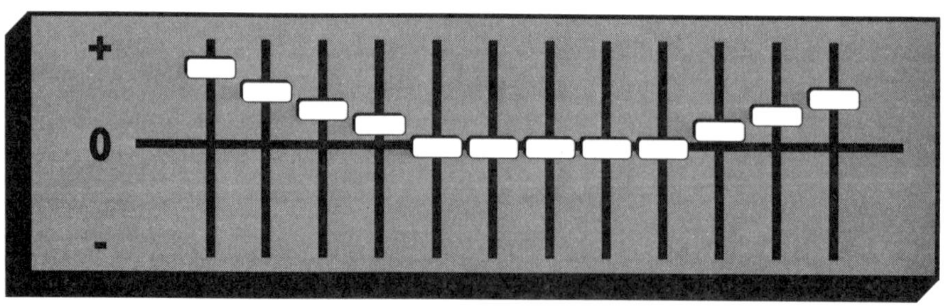

Fig. 3-12: The sliders on a graphic equalizer give a "graphic" representation of the frequency response. In this example, the low and high ends are boosted.

The more bands a graphic EQ offers, the more precise the adjustments you can make. A five-band graphic, for example, is good for general sound-shaping; 12 or more bands let you boost or cut specific frequency ranges with much more accuracy. Top-of-the-line graphics provide a band every third of an octave, although you will seldom find this degree of precision in a multieffects.

A *parametric equalizer* is a highly sophisticated form of tone control. Unlike the graphic equalizer, which can boost/cut only at specific fixed frequencies, a parametric can boost or cut over a continuously variable range of frequencies. In addition, the *bandwidth* (the range of frequencies affected by the boosting or cutting) is variable, from broad to sharp (Fig. 3-13). Note that there are also *quasi-parametric* (also called *pseudo-parametric)* equalizers that include frequency and boost/cut controls but no bandwidth control.

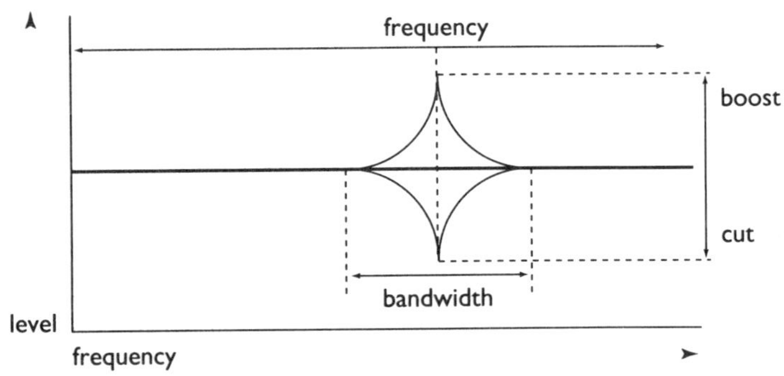

Fig. 3-13: Graphic representation of parametric equalizer parameters.

The typical parametric equalizer includes from one to four individual filter stages. The more stages, the more you can process the sound. For example, with a two-stage parametric the first stage could add a bass boost to the signal, while the second stage adds a midrange notch. Adding another stage would also let you process the high frequencies. Four stage parametrics are quite common and are capable of very precise tone control.

Some multieffects also include specialized types of equalizers, such as software emulations of the passive equalizers found in "vintage" amps and signal processors.

Typical Graphic Equalizer Parameters

Boost/cut determines whether the response at a particular band will be boosted (increased) or cut (decreased). Each band will have its own boost/cut parameter.

Center frequency. While not a common feature, a few graphics now let you vary each band's frequency over a small range. The reason for doing this is because in some cases, the frequency that you might want to boost or cut may not occur exactly where the bandpass filter is located (what happens when you want a boost at 1.5 kHz, and the nearest filters are located at 1 kHz and 2 kHz?). Generally, the available frequency variation will not be too great; if you need more precision in choosing a frequency, use a parametric equalizer.

Typical Parametric Equalizer Parameters

Boost/cut determines whether the frequency response will be boosted (emphasized) or cut (de-emphasized) at the part of the audio spectrum selected by the frequency control.

Frequency sets the specific part of the audio spectrum where the boosting or cutting occurs.

Bandwidth, resonance, or *Q*. This control determines the sharpness of the boosting or cutting action. Narrow bandwidth settings affect a very small part of the audio spectrum, while broad bandwidth settings process a broader range.

Equalizer Tips

Here are some suggestions on applying equalization.

- **Creating new sonic personalities.** EQ can change a sound's character—for example, turn a brash rock piano sound into something more classical. This type of application requires relatively gentle EQ, possibly at several different frequencies; a graphic equalizer works well. Parametric EQs may not have enough bands to affect all the desired sections of the audio spectrum.

  Musicians often summarize an instrument's character with various terms, both positive and negative. Following is a *very* subjective interpretation of which frequencies correspond to these terms.

Range	Positive Correlation	Negative Correlation
20 Hz - 200 Hz	Bottom, depth	Boomy
200 Hz - 500 Hz	Warmth, dark	Muddy
500 Hz - 1.5 kHz	Definition	Honk or nasal
1.5 kHz - 4 kHz	Articulation, "snap," presence	Strident
4 kHz - 10 kHz	Bright, higher range of presence	Screechy, thin
10 kHz - 20 kHz	Sheen, transparency	Thin

  For example, to increase warmth, apply a gentle boost (3 dB or so) somewhere in the 200-500 Hz range. On the other hand, if you consider a sound muddy, try cutting the response in the same region.

- **Cutting is often better than boosting.** The more you boost, the more likely you'll overload the module and cause distortion. If you want to boost the bass and treble, you can accomplish the same result by cutting the midrange instead. This will still make the bass and treble seem more prominent, yet reduce the chance of distortion.

- **Making comparisons.** Use the bypass switch to constantly compare the equalized and non-equalized sounds. You don't want to get into a situation where you boost the treble a lot, which makes the bass seems thin so you boost that, which then makes the midrange seem weak so you boost that, and so on until everything is turned up to 11 (good for Spinal Tap; bad for more successful musicians). Always use the minimum amount of equalization necessary. Just a few dB of change can make a big difference to the sound.

- **Better lead guitar tone for a better mix.** Suppose you're recording a band, and it's time to mix down the tracks into a master tape. You bring up the lead guitar, but it just doesn't seem to "cut." So you raise the level a bit more, but now the guitar drowns everything out.

  This happens because the guitar's frequency range overlaps with a lot of the other instruments. With EQ, we can accent a portion of the guitar to make it real noticeable, but not bring up the entire guitar sound, thus leaving space for the other instruments.

  Adding a bit of a peak around 3 to 4 kHz really makes a guitar solo stand out. Since that's above the range of the toms, bass, and most rhythm-oriented keyboard parts, there's little interference with these instruments. So, the guitar comes through loud and clear, but doesn't step on anything else.

- **Equalizing piano for better vocal support.** Suppose you're playing a rhythmic piano part behind a vocalist, but since the piano and voice occupy a similar frequency range, they conflict. The solution: pull back on the piano's midrange somewhat to make room for the vocal frequencies. You can then mix the piano higher in level, yet still not get in the way of the vocals. This technique also works well with rhythm guitar.

- **Eliminating hum.** A parametric equalizer is an excellent way to reduce hum. Simply set the equalizer for maximum cut and narrowest bandwidth, then dial in 60 Hz (you'll know you're at the right frequency because the hum will disappear). However, if the hum generates harmonics, you'll need additional stages to notch those out as well.

- **Reducing noise with older effects.** If you're fond of some favorite old effect but wish it had less noise, try using a parametric to reduce frequencies in the range of 10 kHz to 20 kHz.

- **Making distortion more responsive.** Putting EQ before distortion can make an instrument seem more "touch-sensitive." This is because distortion usually affects all frequencies more or less equally. By gently boosting the midrange over a fairly broad range *(e.g.,* 200 Hz to 1 kHz) prior to distortion, the notes you play in this range will distort at lower levels, which makes the distortion seem more responsive.

- **Fixing dead spots on bass.** Basses (and guitars) sometimes have "dead spots" on the neck that don't quite seem to have the same power as the other notes; this is a job for parametric equalization. Turn the volume down on your amp, turn the boost and bandwidth controls up full, and play the dead note repeatedly while sweeping the parametric's frequency control. When the parametric hits the right frequency, the note will jump out (in a possibly obnoxious fashion, which is why you should turn down the amp first). Once the frequency control is set properly, reduce the amount of boost until the dead note is the same level as the other notes. If the note sounds too "peaky," reduce the bandwidth control as well.

- **Emphasizing and/or differentiating instruments.** Finding and cutting specific frequencies can eliminate "fighting" between competing instruments. For example, suppose you're mixing a tape track (or live concert) and there are two woodwind parts with resonant acoustic peaks around the same frequency. When playing together they gang up on that part of the frequency spectrum, which makes them difficult to differentiate. Here's a way to work around this:

  1. Find, then reduce, the peak on one of the instruments. To do this, set a parametric EQ for lots of boost (10-12 dB) and fairly narrow bandwidth (around a quarter-octave or so). As the instrument plays, slowly sweep the frequency control. Any peaks will jump out due to the boosting and narrow bandwidth; some peaks may even distort. Find the loudest peak, and cut at that frequency until the peak falls into balance with the rest of the instrument sound. You may need to widen the bandwidth a bit if the peak is broad.

  2. Note the amount of cut and bandwidth that was applied to reduce the peak.

  3. Using a second stage of EQ, apply a roughly equal and opposite *boost* at either a slightly higher or slightly lower frequency than the natural peak.

  Both instruments will now sound more articulated, and because each peaks in a different part of the spectrum, they will tend not to interfere with each other.

- **Using EQ when mixing in the studio.** Wait to apply most EQ until you start setting levels; remember, EQ is all about changing levels—albeit in specific frequency ranges. Any EQ changes will alter the overall balance of instruments. Another reason for waiting a bit is that instruments equalized in isolation to sound great may not sound all that wonderful when combined. If every track is equalized to leap out at you, there's no room left for a track to "breathe."

Time Delay: Flanging, Chorus, Echo

A variety of effects are based on time delay. These include flanging, echo, chorusing, tapped delay, and even esoteric functions like stereo generation from a mono signal, or "psycho-acoustic" panning *(i.e.,* you're not panning the signal conventionally with level changes, but it sounds like it).

Some multieffects include dedicated effects for each function; others simply include a general purpose time delay effect that is flexible enough to provide these different options. In any event, all time delay-based effects have many common parameters.

Understanding Time Delay

Fig. 3-14 shows the block diagram for a typical mono time delay effect. In addition to delaying the signal being processed, there will also be a parameter to mix the straight and delayed signals, and usually, the option to feed some of the delayed signal back to the input (called *regeneration, recirculation,* or *feedback).* With echo, listening to the delayed output without any regeneration produces a single echo. Adding regeneration means that when the echo appears at the output, it also feeds back to the input. Once this regenerated signal passes through the time-delaying circuitry, it creates a second echo; continuing to feed back more of the output produces more echoes.

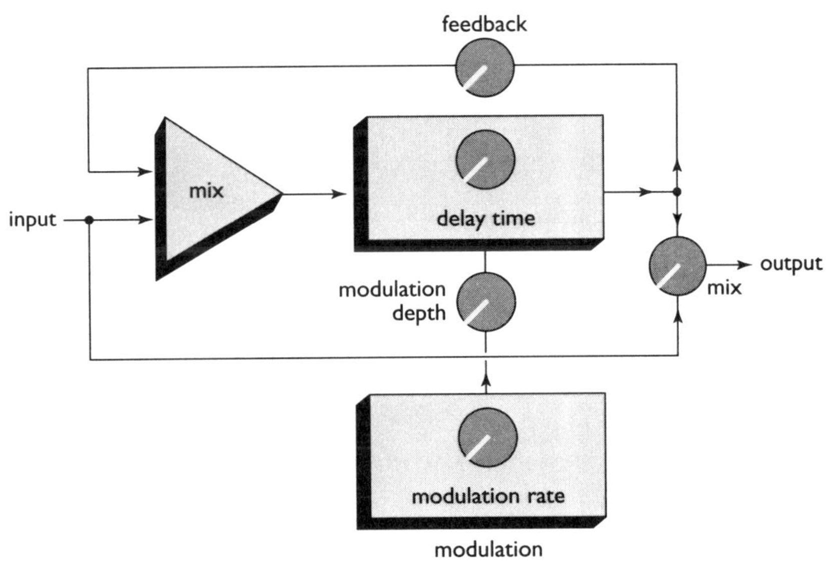

Fig. 3-14: Block diagram of a mono delay line effect. In a multieffects this is all implemented in software, but the functions emulate traditional delay line hardware. A stereo delay will include an additional set of controls for the second channel.

Modulation, a regular (or sometimes random) varying of the delay time over a particular range, is another key element of time delay-based effects. These variations produce an animated kind of sound as the delay time sweeps back and forth between a maximum and minimum value. Modulation parameters include modulation *speed* or *rate* (how fast the variations occur) and *depth* or *intensity* (the spread between the minimum and maximum delay times).

Tapped delays offer several delay lines, each with adjustable delay time and feedback (Fig. 3-15). Some also provide individual modulation for the different taps. This can make a more complex sound than simple stereo or mono delay lines.

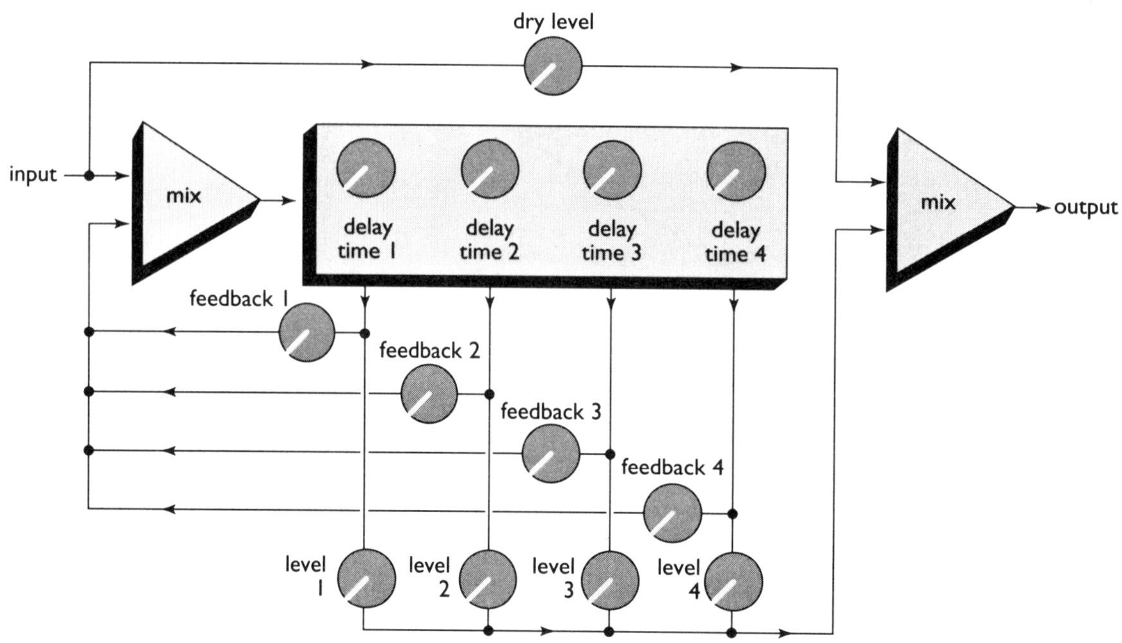

Fig. 3-15: A tapped mono delay line with four taps. Each tap has adjustable delay time, feedback, and output control.

The Time-Shift Spectrum

Different amounts of time delay give different effects. While there is no standard definition of the time-shift spectrum, here are some guidelines as to what sounds are associated with various delay times. These delays are so short that they're given in milliseconds (1/1000 of a second, abbreviated ms).

- **0 to 15 ms delays.** Mixing a signal delayed by 0 to 15 ms with an equal amount of non-delayed signal produces *flanging,* a dramatic special effect which imparts a "jet airplane" sound to the signal going through the flanger.

- **10 to 25 ms delays.** Mixing a signal delayed by 10 to 25 ms with a non-delayed signal produces a *chorus* effect. This creates a full, animated sound which resembles the sound of two instruments playing at once. Modulating (varying) the delayed signal's delay time adds "motion."

- **25 to 50 ms delays.** This starts crossing over into the echo range, where you can perceive that the delayed signal is occurring later in time with respect to the non-delayed signal (with flanging and chorusing effects, it's difficult to tell that an actual delay is taking place because the delay time is so short). 25 to 50 ms delays yield what is popularly called "slapback" echo, a very tight echo sound.

- **50 ms and up.** This is the range covered by most echo units. Longer delays (say, over 250 ms) give a spacey, emotional sound, while shorter delays give more of a doubling effect, or the effect associated with playing an instrument in a small room with hard surfaces.

Typical Time Delay Parameters

Initial delay sets the amount of delay time. With echo, this is the time interval between the straight sound and the appearance of the first echo. With flanging and chorusing, modulation occurs around this initial time delay. Stereo devices usually provide independent delay times for the left and right channels.

Some multieffects let you synchronize the delay time to MIDI clocks (see Chapter 5 on MIDI for more information). Providing that MIDI clock messages set the song's tempo, this feature locks the multieffects' delay time to the tempo. Another option is a tap function, where hitting a switch or button twice sets the delay time interval (in other words, this is the time between the first and second taps).

Balance, mix, or *blend.* Since you rarely want to hear the sound of the delayed signal by itself, this parameter adjusts the balance between the straight and delayed signals. With flanging, you generally set this control for an equal blend of straight and delayed signals; chorusing typically uses more straight than delayed sound. With slapback echo, the delayed sound will tend to be mixed further back. With longer echoes, the amount of delayed sound will depend on how "murky" a sound you want: increasing the amount of delayed sound gives a swimming-in-echo effect, while adding in only a little delayed sound provides more of an ambience effect.

Feedback, recirculation, or *regeneration.* This parameter determines how much of the output feeds back into the input. With echo, minimum feedback gives a single echo; increasing this parameter increases the number of echoes.

With flanging, introducing feedback increases the effect's sharpness, much like increasing the resonance control on a filter. The effect with chorusing is similar. However, dedicated chorus effects may not include a feedback parameter because adding feedback makes the sound more "unnatural."

Feedback phase. This parameter is most relevant to flanging but can also be useful with chorusing. With in-phase (positive) feedback, the flanged sound is metallic and "zingy" since it accentuates positive harmonics. Out-of-phase (negative) feedback emphasizes odd harmonics, which produces a hollower, "whooshing" sound. With longer delays, out-of-phase feedback can sometimes interact with your straight signal in such a way that it weakens the straight signal sound.

There are several ways to implement this function. One option is a phase "switch" parameter that selects positive or negative phase. Another is to calibrate the feedback parameter in positive or negative numbers *(e.g.,* 0 to +100 varies the amount of positive feedback from minimum to maximum, while 0 to -100 similarly varies the amount of negative feedback).

Feedback tone controls. Some musicians prefer the sound of tape echo instead of digital echo, probably because tape's inferior high frequency means that as each echo recirculates through the delay line, it loses more and more of its high frequency edge. As a result, the early echoes never "step on" the straight sound or on newer echoes. With a digital delay, echoes become progressively softer in terms of level, but their frequency response doesn't necessarily change. Psycho-acoustically, this sounds unnatural since in a traditional acoustic space, echoes indeed lose higher frequencies faster than lower frequencies. So, many digital delay effects include a parameter to restrict the feedback path's high frequency bandwidth.

Sweep range, modulation amount, or *depth* determines how much the modulation section (also called LFO, or sweep) varies the delay time. For example, a delay with a 2:1 sweep range could sweep over a two-to-one time interval *(e.g.,* 5 ms to 10 ms, or 100 ms to 200 ms). Practically speaking, a wide sweep range is most important for dramatic flanging effects—chorus and echo don't need much sweep range to be effective. With longer delays, adding a little bit of modulation can give chorus-like sounds, but too much modulation will cause detuning effects.

With chorusing, the main use for depth is to strike the correct balance between detuning and chorusing; too little depth gives a weak chorusing sound, whereas too much depth sounds out-of-tune.

Modulation type. Fig. 3-16 shows the most popular forms of modulation.

- *Triangle wave* varies the delay time smoothly from a maximum to minimum value in a cyclical manner, and is most often used with flanging and chorusing.

- *Sine wave* sounds very similar to the triangle wave but can be a little bit "smoother."

- *Square wave* switches cyclically between two delay times.

- *Logarithmic* is particularly useful with flanging. It stays for a longer time at the top end of the sweep, then dips down dramatically into the lower range.

- *Exponential* is another flanging-oriented waveform that stays for a proportionately longer amount of time at the lower end of the sweep range, then sweeps up rapidly to the peak.

- *Random* changes delay time values at random.

- *Smoothed random* is like random, but rounds off the edges.

- *Envelope follower* is not a periodic waveform, but causes the delay time to vary in response to the dynamics of the instrument's signal (for more information on envelope followers, see the section on Envelope Controlled Filters). This means that the delay time variations always relate rhythmically to your playing, rather than being controlled by a modulation source (such as the triangle wave) which is seldom in sync with the song's tempo. Envelope followers are less common than the types of modulation mentioned above, but can work very well with flanging and chorusing effects.

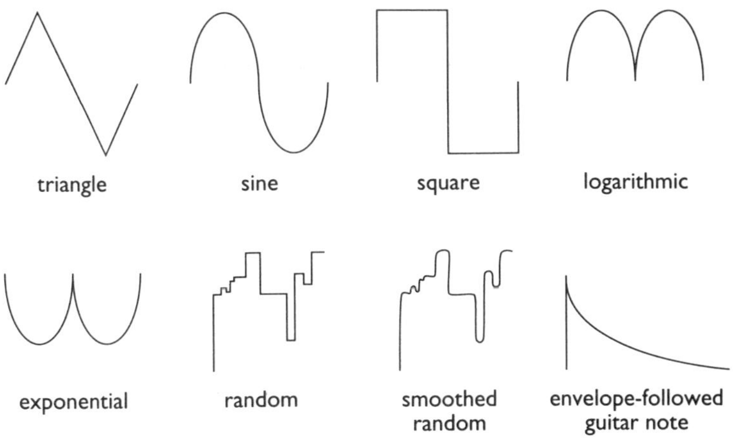

Fig. 3-16: The most common types of modulation signals.

Modulation rate sets the modulation frequency. Typical rates are 0.1 Hz (1 cycle every 10 seconds) to 20 Hz. With flanging and chorusing, modulation causes the original pitch to go flat to a point of maximum flatness, return to the original pitch, go sharp to a point of maximum sharpness, then return back to the original pitch and start the cycle all over again. A slower rate produces a slow, gradual detuning that gives a majestic, rolling chorusing or flanging sound. Faster rates produce a more "bubbly" effect.

Note that the rate parameter interacts with the depth parameter because the total amount of pitch change depends not just on the amount of pitch change, but also on the rate. For example, combining full depth with a fast rate setting can sound out-of-tune, whereas the same amount of depth coupled with a slow rate sounds just fine.

Hold or *"freeze."* Sometimes this is a separate function, but some delays let you capture (sample) a sound in the delay's memory and repeat that captured section of sound indefinitely. If you're into electronic sounds and special effects, hold is definitely a useful option. Typical parameters associated with this feature are:

- *Pitch.* Sometimes the delay control can vary the pitch of the stored sound.

- *Record trigger threshold.* This sets a level; the sample to be recorded must exceed this level to begin the recording process. This helps eliminate "dead air" at the beginning of the sample.

- *Sample start point* selects the point within the sample where, upon receiving a trigger, playback begins.

- *Sample stop* or *end* point selects the point within the sample where playback ends.

- *Playback trigger mode* sets whether the sample plays back via a manual or audio trigger (if available). For example, if there's a really great-sounding snare drum sound stored in memory, you could use the audio trigger function to trigger the great-sounding snare from a lousy-sounding snare recorded on tape.

- *Loop mode.* Generally you'll have two choices (Fig. 3-17). With loop on, once the sound reaches the loop end point it jumps back to the loop start point and continues playing *ad infinitum.* With loop off (also called *one-shot* mode), the sampled sound simply plays through from beginning to end.

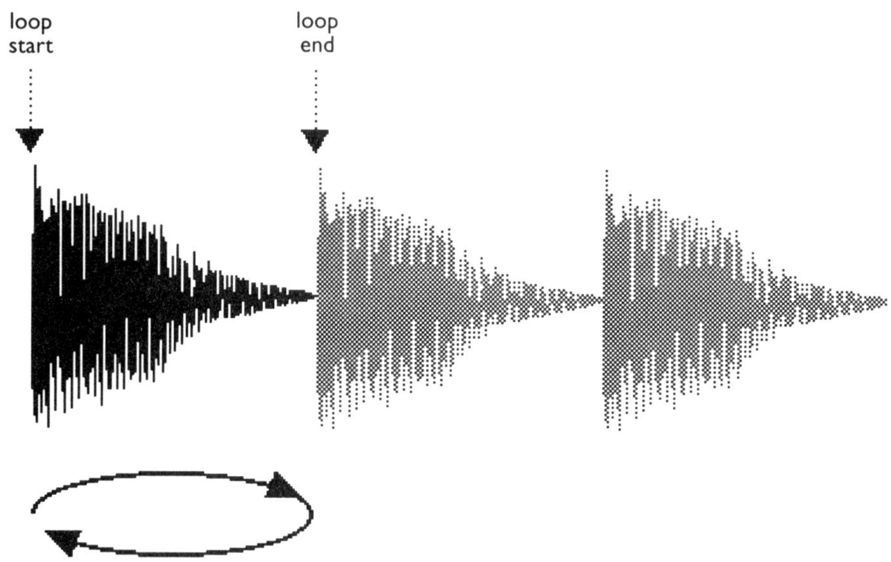

Fig. 3-17: In one-shot mode, only the portion of the sound shown in black would play. With loop on, this portion would repeat over and over again, as represented by the additional sounds shown in gray.

Long freeze times can create solid-state tape loop effects. Shorter loops are good for special effects and generating unusual tones, but tend to be not as musically useful (in a conventional sense) as longer loops.

Time Delay Tips

Time delay-based functions can give a rich repertoire of sounds. Here are some suggested parameter values to get you started.

- **Vibrato.** Set a short initial delay (5 ms or so), listen to the delayed sound *only,* and modulate the delay with a triangle or sine wave at a 5 to 14 Hz rate. The greater the amount of modulation, the deeper the vibrato effect.

- **Positive flanging.** Adjust the mix for equal amounts of delayed and straight signals, set the initial delay for 5 to 10 ms, select positive feedback (even harmonics), and add slow, deep modulation. More regeneration gives a sharp, metallic sound.

- **Negative flanging.** Use the same basic settings as positive flanging, but apply negative feedback (odd harmonics). This produces a more "whooshing" type of timbre.

- **Rotating speaker simulation.** For best results, set the initial delay in the 5 to 15 ms range, add modulation sparingly, set the blend control for slightly less delayed sound than straight sound, and don't use any recirculation. The rotating speaker simulation works best at faster modulation speeds.

- **"Machine Gun" reverb.** There's a zone between 15 ms (where flanging turns into chorusing) and 50 ms (a typical minimum delay for echo) where the echoes are very tightly spaced. Increasing the amount of feedback produces a series of very tight echoes; the sounds blend together into a pulsing, reverb-like effect.

- **Chorusing.** This turns six string guitars into pseudo-twelve strings, thickens vocals, and makes reverb sound great. Set an initial delay around 20 ms, use no feedback, adjust the mix control for an equal balance of straight and delayed sounds, and inject a subtle amount of (preferably smoothed random) modulation. Lengthen the delay time or increase the modulation depth to increase the effect.

- **Phase shifter simulation.** Phase shifters, introduced in the early 70s, were intended to simulate the sound of tape flanging. However, since these devices did not use true time delay technology, they produced a subtler effect than flanging. Although flanging and phase shifting sound quite different, it's possible to simulate a typical phase shifter sound with a multieffects chorus or flanging module.

  The control settings are similar to flanging, but this application sweeps a much narrower delay range. Set the initial delay for approximately 2 ms; also select minimum feedback, a fair amount of modulation at a medium rate, and an output mix that's 50% straight and 50% delayed sound. Optimize the sound by playing with the initial delay and modulation width parameters. For a sharper sound, turn up feedback somewhat (negative feedback seems to give the most realistic phase shifter sound).

- **Tone control (comb filter).** Mixing a straight signal with the same signal passing through a short, fixed (unmodulated) delay produces a filtering effect that changes the signal's tone. This type of filter is called a "comb filter" since plotting its frequency response looks somewhat like a comb—there are lots of peaks and valleys. Suggested control settings are an initial delay of 0 to 10 ms, minimum feedback, no modulation, and an equal blend of processed and straight sound (this produces the sharpest peaks). Increase the amount of feedback to increase the depth of the filter peaks and dips; this creates a more "resonant" sound. For a variation on this timbre, change the feedback phase from positive to negative to reverse the positions of the frequency response peaks and valleys.

- **Automatic Double Tracking (ADT).** This simulates the effect of playing a part then overdubbing a second part to give a thicker sound. Set the initial delay for a short echo (around 30 to 40 ms). Adding a very small amount of modulation (preferably random) alters the delay time to better simulate true double tracking, where small timing inconsistencies keep each track from being identical. While the effect is conceptually similar to chorusing, chorusing is more closely related to flanging, while ADT is more closely related to echo.

- **50s-style slapback echo.** In the early days of recording, echo effects were provided by tape recorders rather than by the digital wonder boxes we know today. Feeding a signal into the tape recorder input, going into record, rolling tape, and monitoring the signal coming from the playback head produced a typical delay of around 70 ms. Select a delay program, and set the delay time to 70 ms to recreate the vintage echo sound; add feedback to taste (use positive feedback) and minimum modulation.

- **Mono to pseudo-stereo conversion.** With stereo chorus (and flanging), set the chorus depth to maximum and rate to minimum (or off if possible). This creates a stereo spread without the motion that would result from having a higher modulation rate.

- **"Canon" echo.** A canon is a type of musical composition that plays the same melody by two or more parts (probably the best-known examples are "Row Row Row Your Boat" and "Frère Jacques"). A multieffects can give a simplified version of this effect by delaying what you play, then repeating it a fraction of a second later. Set the delay time for the desired time interval between when the first part plays and when you want the second part to start playing; minimum feedback creates a single repeat. Use no modulation, and set the output mix for 50% straight and 50% processed sound.

Relating Echo Times to Song Tempos

Setting the echo repeat time to equal a particular rhythmic value, such as an eighth or quarter note, "synchronizes" the echo time to the tempo (synchronizing to MIDI clocks does this automatically, but only a few multieffects incorporate this feature). Here's a simple formula that translates beats per minute (tempo) into milliseconds per beat (echo time):

$$60{,}000/\text{tempo} = \text{time (in ms)}$$

For example, if a tune's tempo is 120 beats per minute, then the number of milliseconds per beat is:

$$60{,}000/120 = 500 \text{ ms}$$

So, setting the echo for 500 ms gives an echo every beat. Repeatedly dividing the delay time by two (*e.g.*, 250 ms, 125 ms, 62 ms, and 31 ms) gives progressively tighter echoes that remain in sync with the tempo. Similarly, multiplying by two gives longer echo intervals. (By the way, multieffects calibrations may vary somewhat, so treat the displayed value as a close approximation. It might be necessary to tweak the delay time a bit to have it sync up perfectly with your music.)

If you don't know the tempo of a piece of music, there are two ways to determine the proper echo time. Method #1 is simple: grab a stopwatch, count the number of beats in a minute, then apply the above formula.

Method #2 is to turn up the echo feedback a substantial amount—enough to generate, say, seven or eight echoes. Next, set the delay time somewhere in the 250 ms to 500 ms range. While the music plays, hit a staccato chord and listen carefully to the echoes. If they lead the beat, increase the echo time; if they lag the beat, decrease the echo time. Eventually, through trial and error, you'll find a delay time setting where each echo occurs on the beat, at which point you can pull back a bit on the feedback (if desired). As with the above example, divide the number of milliseconds by two, four, eight, etc. for shorter echoes, or multiply by powers of two for longer echoes.

Stereo echo allows for even more options, such as polyrhythmic effects and panned echoes that add a real spaciousness to the sound. Try setting one channel as described above, then adjust the other channel's echo time to some triplet value—say, 66% or 33% of the first channel's delay time. This sounds so great you have to hear it to believe it.

Long Delay Applications

As memory becomes less expensive, available echo times get longer. Older multieffects were lucky to get 1 second of total delay, but the trend is toward the ability to create longer and longer echoes if desired. Usually devices with lots of memory will also let you "freeze" (sample) sounds and repeat them indefinitely, as described earlier in this chapter.

Long delays open up several new types of applications. Consider tape loops, once a popular technique in avant-garde electronic music. To make a tape loop, you record a sound (usually of short duration) and splice the end of the taped sound back to its beginning. Thus, as you play the loop, the sound repeats over and over and over and over and over and over and over and...you get the idea.

If you recorded a short sound—for instance, 500 milliseconds of a printing press—and looped that, you would have a 120 BPM printing press-based rhythm track. The only problem with this technique is that tape recorders were not intended to play back loops, which created a variety of mostly mechanically-related problems.

A long delay with a freeze feature lets you create instant tape loops: play a sound into the delay, then freeze and repeat the sound. Unlike a tape loop system, there are no moving parts; the only disadvantage is that the sound will be lost as soon as the AC power goes away (the memory used in multieffects generally cannot store sounds permanently).

Long echoes also have their uses, particularly for the "Frippertronic" music popularized by guitarist Robert Fripp. This effect is a refinement of an electronic music technique that involved using two tape recorders to create extremely long echoes. The two recorders were set up side-by-side, often separated by several feet, with the tape being supplied from the supply reel on the first machine and taken up by the takeup reel on the second machine. The sound would be recorded at the record head of the first machine, travel over to the second machine, and play back from the second machine's playback head. As you might imagine, this technique was not super-reliable, especially if the distance between tape units was substantial. Long electronic delays avoid the associated mechanical problems.

Increasing the feedback parameter value with long echo times produces multiple echoes instead of a single repeat. For example, if you play a note on the first beat of a measure (and the delay time is the same length as a measure), then you will hear the first echo exactly one measure later, a second echo two measures later, a third echo three measures later, and so on.

Another long delay application is live "digital multitrack recording." Combining a long echo time with lots of feedback simulates multitrack recording, since when you play a musical phrase the feedback will cause it to repeat over and over again; you may then "overdub" by playing new lines on top of this repeating phrase. These overdubs will repeat as well, so you can continue overdubbing parts to build up several tracks of sound.

With some multieffects, you can then select freeze to store the multitracked sound. The only limitation is that until you freeze the sound, every time a "track" repeats its fidelity deteriorates somewhat. However, you should be able to record plenty of overdubs before the sound quality becomes objectionable.

Here's an example. If the delay time is set for the equivalent of two measures of music, you could initially play a two-measure bass line on programmable synthesizer, then switch over to another patch and overdub a two-measure melody over the bass line. As the bass and melody continue to repeat, you could then change patches a third time and lay down yet another part. This can continue until the sound quality becomes intolerably poor, or until you've played everything that you wanted to play.

The amount of feedback is critical—it must be turned up high enough so that each successive echo maintains the same volume level as the previous echo, yet not be turned up so high that runaway feedback occurs. To set this control properly, temporarily select a fairly short delay time *(i.e.,* 50 ms) and play a sound through the delay. Set feedback to the maximum possible setting short of runaway feedback, then return to the desired delay setting. Use positive feedback for this application with no modulation. You'll probably want the output mix to be half straight and half processed sound.

Attack Delay

Although not too common a module, attack delays are showing up in more and more multieffects. It's also sometimes possible to have a noise gate mimic the effects of an attack delay unit (see the section on noise gates for more information).

The attack delay unit senses when the input signal (a note or chord) exceeds a certain threshold and fades the signal in gradually, giving effects that resemble backwards tape effects, as well as the attack characteristics associated with woodwind instruments.

There are certain challenges associated with this process. You have to play very cleanly to avoid having false triggers, and the threshold setting is critical—set it too high, and some notes won't trigger the attack delay; set it too low, and just touching the strings will create a fade in.

Typical Attack Delay Parameters

Sensitivity or *threshold* sets the threshold, thereby influencing the trigger characteristics. An attack delay without a sensitivity control can still give workable results, although you will have to play more carefully to match the preset threshold.

Attack time determines how long it takes for the signal to fade up to full volume after triggering the attack delay (Fig. 3-18).

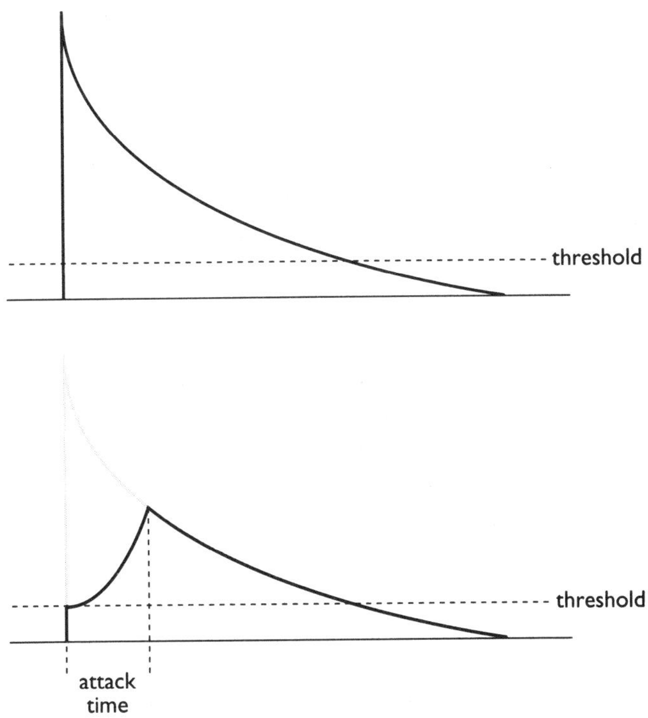

Fig. 3-18: Graphic representation of attack delay parameters. The upper diagram shows the unprocessed signal. The lower diagram shows how adding an attack time cuts off the signal's attack.

Attack Delay Tips

- **Playing style suggestions.** Don't expect an attack delay unit to faithfully follow your playing and trigger perfectly every time you play a note—you will almost certainly have to alter your playing style somewhat (guitarists in particular must pick in a consistent manner) to obtain the best results. You might have to choose between setting a fairly high threshold for chords, or a lower threshold to handle single note lines.

- **Allowing for trigger reset.** Insert a brief pause, or mute the signal between notes, to allow the trigger circuitry time to reset. Only very sophisticated units will allow you to play continuously and still achieve an attack delay effect on each note. Because of this, attack delay effects are generally most useful when playing slowly.

- **Selective attack delay.** Note that having to leave a space between notes can also be used to advantage. For example, suppose you're playing a series of chords and you want the attack delay effect only on the first chord of every series. Simply insert a pause just before the first chord to give the circuitry time to reset, and then play the rest of the chords in a continuous manner to prevent the unit from retriggering.

Envelope Controlled Filter

An envelope controlled filter (also called an envelope followed filter) produces sounds that are similar to a wa-wa pedal. However, rather than having the filter's resonant frequency respond to the position of a pedal, an envelope followed filter responds to the dynamics of your playing. In other words, playing louder produces the same effect as pushing down on a wa-wa pedal (namely, increased treble emphasis), while playing softer produces the same effect as pulling back on the pedal (bassier emphasis).

From an audio standpoint, the filter section is similar to the filter in a wa-wa pedal, but usually includes some enhancements. The range will probably be greater than that of a pedal, and there may be a choice of filter modes (lowpass, bandpass, highpass) and resonance amount.

Typical Envelope Followed Filter Parameters

Envelope followed filter controls must be set carefully to obtain good results. Typical controls include:

Sensitivity determines how much the dynamics affect the filter's resonant frequency, and matches the envelope follower response to your playing characteristics. With maximum sensitivity, it takes very little audio signal to push the filter up to its highest resonant frequency, while with minimum sensitivity, you'll have to really bash on your instrument to make the filter respond. Adjust the sensitivity parameter so that with the peaks of your playing, the filter hits its highest resonant frequency.

Initial frequency sets the filter's initial resonant frequency when there is no audio input. With guitar, setting this to the lowest note (80-100 Hz) provides the maximum sweep range, but setting the initial frequency to higher settings can produce good results. For example, if you adjust the initial frequency to around 2 kHz and sweep the filter upwards from that point, the result is a treble-enhancing effect that varies with the dynamics of your playing. The envelope controlled filter then becomes more of a sophisticated, responsive tone control than simply an imitation wa-wa. This effect can also work very well with vocals to increase intelligibility.

Resonance determines the filter's sharpness, or bite. Low resonance creates subtler, more muted filter sounds, while high resonance creates the "quacking" sounds associated with some keyboard synthesizer patches. Extreme amounts of resonance can create an almost whistling kind of effect which can be musically useful in some applications.

Sweep direction. Normally, most envelope followed filters start at a low resonant frequency, and increase in frequency as you increase the input signal ("wa" sound). By reversing the sweep direction, the filter starts off at a high resonant frequency; increasing the input signal level lowers the frequency ("ow" sound).

Response curve. Some units offer a choice of logarithmic or linear response characteristics. A linear response gives the greatest amount of dynamic range, while a logarithmic response compresses dynamic range somewhat to give less extreme response variations.

Response time determines how closely the filter tracks your dynamics. For example, with a slower response time, the filter frequency will fade up to maximum when you hit a note sharply rather than instantly hit the maximum.

Blend or *mix* determines the balance of the straight and filtered sounds. Any filter removes part of a signal, thereby creating a thinner sound; while this may often be desirable, there are also times where this thinness works to your disadvantage. Solve this by using the mix parameter to increase the percentage of the straight signal.

Envelope Followed Filter Tips

- **Optimum signal chain position.** With multieffects that let you change an effect's position within the signal chain, place the envelope controlled filter before any modules that reduce dynamic range (distortion, compressor, limiter, etc.). Although placing an envelope followed filter after these devices will work, the restricted dynamic range means that the filter's resonant frequency will pretty much be either at maximum or minimum, with few variations in between these two extremes.

- **Optimum parameter values.** Adjust the input level and sensitivity parameters carefully. Not enough level will produce a muffled sound; too much level produces a less dramatic effect. With guitar, you may also need to make some changes at the instrument itself. For example, if a guitar's low strings put out much more signal than the high strings, the envelope follower will respond more to the lower strings. Angling the pickup so that the higher strings are closer to the pickup and the lower strings are further away takes care of this.

Pitch Transposers

Pitch transposers synthesize a parallel harmony line from an input signal. For example, if the transposer is programmed to synthesize a note a major third above the input signal, every note you play will be accompanied by another note a major third higher.

Some pitch transposers go beyond simple parallel harmonies and let you specify a key and mode (major, minor, etc.). As you play, a perfect harmony will follow your signal. Even if your multieffects doesn't have this kind of "intelligent" harmonization, you can often change the transposition amount via MIDI by using continuous controllers as you play. See Chapter 5 on MIDI applications for more information about continuous controllers.

It takes a lot of processing power to do pitch transposition, and the sound sometimes suffers as a result. For example, there might be a fluctuating tremolo effect, or occasional glitches. The greater the degree of transposition, the more objectionable these problems tend to be.

Another problem is that the harmony line will often have a slight delay because the computer has to analyze the incoming signal before it can process it and create the harmony. This can take up to several milliseconds. Despite these limitations, though, pitch transposers are extremely useful for thickening a sound and creating novel effects.

Typical Pitch Tranposer Parameters

Transposition or *harmony pitch.* This sets the harmony line interval, which will be continuously variable in semitones over the range of the unit (typically ±1 or ±2 octaves). In most cases there will also be a fine tuning parameter that further varies the range within ±50 or ±100 cents (±1 semitone).

Blend or *mix.* Like most other delay-related modules, pitch transposers have a blend control that determines the balance of the processed and unprocessed signals.

Feedback, regeneration, or *recirculation* feeds some of the output back to the input, and is usually variable from no regeneration to infinite regeneration. With lots of regeneration, the pitch transposer may go into oscillation and emit a continuous stream of sound, just like delay.

Quality or *tracking* trades off sound quality for tracking accuracy and speed. Generally, slower tracking gives better sound quality. This parameter becomes most important with a large amount of transposition, which tends to degrade sound quality.

Predelay sets the amount of delay before the harmonized signal appears.

Intelligent harmony settings. These usually require that you enter the key and scale in which you're playing. The pitch transposer will then generate harmonies based on the rules of harmony for the scale you've chosen.

Modulation parameters. These let you superimpose vibrato and chorusing effects on the pitch-transposed harmony line.

Pitch Tranposer Tips

- **Glissando.** With the transposed pitch set very slightly higher than normal pitch (a few cents), advancing the regeneration control recirculates and pitch shifts each note, thereby initiating a stepped, upward glissando effect (the harmony pitch control controls the step interval). Setting the transposed pitch slightly lower than the original signal produces a similar downward glissando. The downward glissando effect is particularly dramatic with single note solos, since as each note ends, it sounds like it's spiralling downwards. Another excellent application is with hand claps. By setting the unit for a subtle downwards spiralling, the claps sound bigger-than-life and much thicker than normal.

- **"Bell trees."** Holding a single note with the regeneration advanced, and then varying one of the pitch controls, produces a variety of useful effects—from pseudo "bell trees" (seemingly endless upwards or downwards spiralling) to complex flanging-with-pitch change effects. This effect shows how adding regeneration can change a pitch transposing device into an excellent special effects generator.

- **Expanding the limits of feedback.** Units that offer variable predelay greatly expand what you can do with regenerated sound. For example, delaying the regeneration signal lets you slow down the speed of the glissando mentioned above; with a delay of 0.5 seconds, and a pitch setting one half-step above the original signal, hitting a single note will result in an upward glissando whose pitch increases by one half-step every 0.5 seconds. Also try larger intervals, such as a third or a fifth.

- **Polyphonic octave divider.** Set the transposition interval exactly one octave below the signal being processed for guitar-plus-bass effects. For polyphonic octave multiplication, set the harmony pitch control an octave higher than the signal being processed.

- **Flanging/chorusing effects.** Set the pitch control for a very slight amount of transposition (say, 1/10 to 1/5 of a semitone) and add regeneration to produce some excellent flanging effects.

Noise Gates

Noise gates help remove the noise and hiss produced by signals entering the multieffects, as well as noise generated within the multieffects itself. As long as there is some kind of audio signal going through a noise gate, this "gate" remains open and lets the signal through; but once the signal consists only of noise, the "gate" closes to mute the noise.

To understand the principle of operation, first consider a "manual noise gate." Suppose that you're listening to an audio signal being processed by a relatively noisy effect. As long as the audio signal is present, its level will generally be higher than the noise, thus masking it. However, when the audio signal goes away, the noise is no longer masked and can be audible.

If you now connect a volume control after the noisy effect, turning down the volume whenever there was no audio signal could eliminate the noise. Then, as soon as the audio signal (which masks noise) came back, you could turn the volume up again. However, a manually controlled noise gate cannot react to rapid changes. An electronic noise gate senses the noise/no noise conditions automatically, and closes or opens the gate accordingly.

If the signal at the noise gate input exceeds a certain threshold, the gate opens. If the signal drops below the threshold, the gate closes. Setting the threshold just above the noise level insures that the noise will be muted when no signal is present (Fig. 3-19).

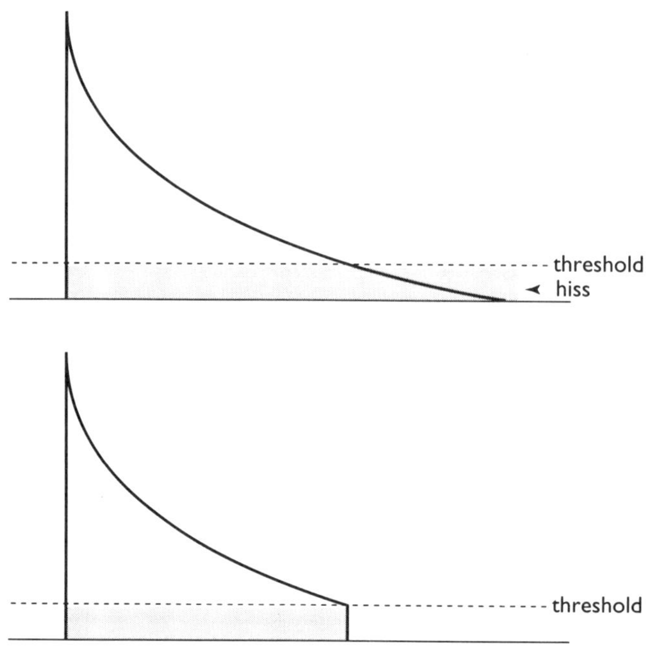

Fig. 3-19: How a noise gate removes noise. The upper diagram shows the signal before gating. In the lower diagram, note that although we've masked the noise while the signal is present, the decay cuts off more abruptly than normal.

Noise gates work best on signals that don't need to be cleaned up too much. If the noise is really severe, eliminating lots of noise also means getting rid of substantial portions of the signal. Another problem occurs with sounds that have long decay times. For example, a guitar's signal level can become erratic towards the end of its decay, and start criss-crossing back and forth across the threshold. This gives a "chattering" type of sound. However, despite these complications noise gates are still extremely useful for keeping overall noise levels down, especially when using a number of effects. As a bonus, some noise gates can also provide special effects.

Typical Noise Gate Parameters

Threshold or *sensitivity* determines the reference level above which the gate opens, and below which the gate closes. Ideally, you should be able to set this reference very low in order to accommodate signals that have very little noise; but higher threshold levels are useful for special effects, such as removing substantial amounts of an instrument's decay to make a more percussive or gated sound.

Attenuation. Some noise gates feature adjustable attenuation in case the contrast between the gate on and gate off conditions is too severe. With less attenuation, the gate doesn't shut down all the way so that some of the signal can still pass through.

Hysteresis affects the way the threshold works. With no hysteresis, the gate opens and closes at the same threshold level. Increasing hysteresis widens the difference between the gate on and gate off levels so that a signal turns on at a higher level than the threshold, and turns off at a lower level than the threshold. This can help minimize false triggering since the input signal has to be at a somewhat higher level than usual to open the gate.

Decay time prevents the gate from closing down abruptly; instead, when the signal goes under the threshold, a certain amount of time elapses over which the noise gate fades out (Fig. 3-20).

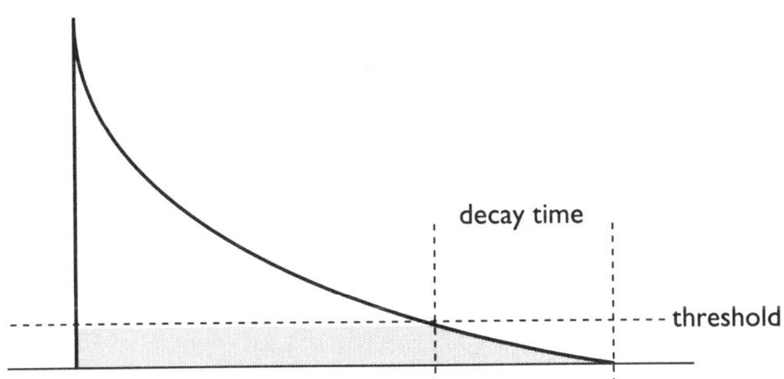

Fig. 3-20: How decay time affects the gated signal.

Attack time works in reverse: when a signal exceeds the threshold, the noise gate fades in over a specified period of time. With long attack times, it may even be possible to simulate an attack delay unit (see the Attack Delay section for more information on this effect).

Reverberation and Gated Reverberation

Reverb simulates the sound of playing in a large hall or auditorium.

A room's acoustical qualities create a characteristic "ambience." The audience hears not only the sound that comes directly from an instrument or amplifier, but myriad sound waves reflecting off the room's walls, ceiling, and floor. What makes the overall sound even more complex is that these waves continue to bounce around the room many times, creating additional delayed reflections. As a result, what the audience hears is a composite of the original audio signal, the first reflections from various surfaces, and the delayed reflections. Eventually these sound waves lose their energy, and the sounds become so weak as to be inaudible.

What's in the room also affects the sound. A thickly carpeted room will easily absorb reflected waves, while a room with extremely hard surfaces will tend to keep the sounds bouncing around. Also, high frequencies are more prone to absorption than lower frequencies.

For centuries, the human ear has become accustomed to hearing acoustical instruments in some kind of acoustic space, and as a result music doesn't quite sound "real" unless there is a degree of ambience. Digital reverberation allows you to precisely program such factors as decay time and number of initial reflections with a great deal of precision, giving you immense flexibility in re-creating the sounds of various acoustical spaces.

Digital reverb can also create spaces that don't exist in nature, such as *gated reverb*. This does the equivalent of adding a noise gate to the end of the reverb, allowing for an abrupt cutoff of the reverb tail. This effect is used a lot with drums.

Typical Reverb Unit Parameters

Fig. 3-21 graphically illustrates reverb parameters.

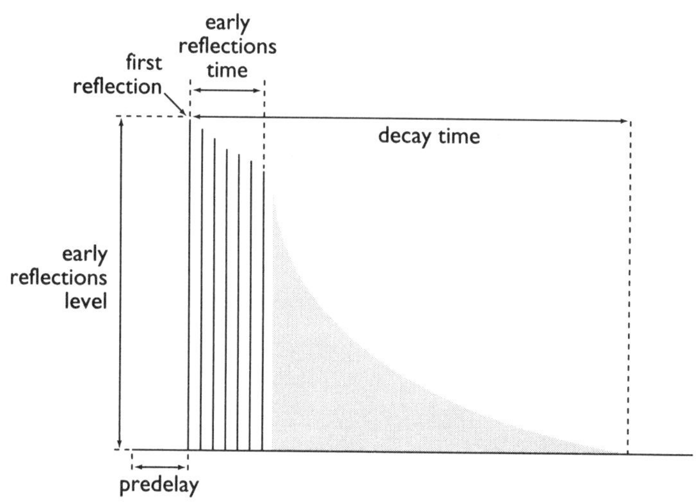

Fig. 3-21: Common reverb parameters.

Type determines the kind of reverb to be emulated: room, hall, plate (a bright, clean type of older reverb unit used in recording studios), spring (the classic "twangy" reverb sound used in guitar amps), etc.

Room size determines the apparent volume of the room (this parameter is sometimes calibrated in cubic feet or meters). Changing this parameter will often change other parameters, such as low and/or high frequency decay.

Early reflections predelay sets the amount of time before the first group of reflections begins, and is usually less than 100 ms (20-50 ms are typical values). A longer predelay setting gives the feeling of a larger acoustical space.

Reverb predelay controls the amount of time before the room reverb sound begins. As with early reflections predelay, this is usually 100 ms or less. Note that reverbs sometimes do not include separate predelay settings for early reflections and reverb, but offer a combined setting.

Early reflections level. Early reflections are closely spaced discrete echoes, as opposed to the later "wash" of sound that constitutes the tail of the reverb. This parameter determines the level of these discrete echoes.

Early reflections shape imparts an envelope to the early reflections. The envelope may attack instantly then decay slowly, build up over time then decay, take a long time to attack then decay abruptly, etc.

Decay time sets how long it takes for the reverb tail to decay to the point of inaudibility. Note that there may be separate decay times for higher and lower frequencies (and possibly midrange as well) so you can more precisely tailor the room's characteristics.

Crossover frequency applies only to units with separate decay times for high and low frequencies. This parameter determines the "dividing line" between the highs and lows. For example, setting the crossover frequency to 1 kHz means that frequencies below 1 kHz will be subject to the low frequency decay time, while frequencies above 1 kHz will be subject to the high frequency decay time. Note that really fancy reverbs may have separate high, mid, and low frequency crossover frequencies. Each of these may have a level control as well.

High frequency rolloff. In a natural reverberant space, high frequencies tend to dissipate more rapidly than lows. High frequency rolloff helps simulate this effect.

Mix, balance, or *blend.* This determines the mix between the reverberated and straight signals.

Reverb density or *spread* determines the space between the first reflection and subsequent reflections. With lower density settings, the first reflection is audible as a separate event, followed by the remaining reflections. Higher density settings move the remaining reflections closer to the first reflection, so that the first reflection joins the overall "wash" of reverb.

Early reflections diffusion is a "smoothness/thickness" parameter. Increasing diffusion packs the early reflections closer together, giving a thicker sound. Decreasing diffusion spreads the early reflections further apart.

Reverb diffusion is similar to early reflections diffusion, but affects the room reverb sound. Some simpler reverbs combine diffusion for the early reflections and the overall reverb into a single parameter.

Gated reverb parameters vary considerably from manufacturer to manufacturer; some multieffects may even include gated reverb as a separate effect. Typical parameters are gate *threshold* (how much of the reverb tail is cut off), gate *shape* (even *reverse* or *backwards reverbs* are possible, where the tail increases to a certain level instead of fading out), and others. It's best to consult the manual for more information on these types of effects since they vary so much among different units.

Note: there is some disagreement among manufacturers as to the exact definition of "diffusion" and "density," so your reverb may function somewhat differently than described above.

Reverb Tips

- **Different instruments can sound better with different reverb settings.** For example, low density settings can be problematic with percussive sounds, since the first reflection could sound more like a discrete echo than part of the reverb. Increasing the density solves this. However, low density settings can work very well with voice to add more fullness to the overall sound.

- **Using "dedicated" reverb.** Because high-quality reverb requires a tremendous amount of computational power, some multieffects include an option that dedicates all the DSP to making reverb, but allows for no other effects. These multieffects will usually offer a "scaled-down" reverb module that does allow for using other effects, but the sound will be compromised compared to the dedicated, reverb-only mode.

- **Altering timbre.** To create a "bigger" sound, set the low frequency decay longer than the high frequency decay. For a more ethereal sound, do the reverse.

Miscellaneous Effects

In addition to the common effects mentioned so far, you may also find some of these more unusual processing modules.

Speaker Emulator

Much of a guitar amp's characteristic sound depends on how the speaker and cabinet shape frequency response. A speaker emulator provides the equalization needed to approximate a guitar amp. Sometimes this will be a multiband equalizer so you can come up with a wide variety of settings; with other multieffects, you'll have the option to choose among several typical frequency response characteristics (four 12" speakers, dual 10" speakers, etc.).

Tremolo

This provides a periodic amplitude change so that the sound seems to "pulsate." A modulation source, such as a triangle or sine wave, controls these amplitude changes (for more information on modulation, see the section on Time Delay parameters). There will usually be parameters for modulation rate, modulation depth, and modulation waveform (if multiple waveforms are available).

Auto Pan

Panning sweeps a signal between the left and right channels in the stereo field. Auto panning automates this so that your signal is constantly shifting between the left and right channels. Generally, this is done with a low frequency oscillator that provides modulation such as triangle wave, square wave, random, etc. There will usually be parameters for modulation rate and depth.

In some multieffects, auto pan may be envelope triggered so that panning occurs when you play a new note or chord. For more information on envelope triggering, see the section on Envelope Controlled Filter parameters.

Note that some auto pan modules can provide tremolo if you monitor only one channel, since the signal "drops out" when it pans over to the other channel.

Ring Modulation

Ring modulation provides bizarre, non-harmonic sounds by modulating one audio signal with another audio signal. Technically, the resulting signal is the sum and difference of the two input signals, with the original signals suppressed. For example, if you feed in a 2 kHz sine wave and modulate it with a 3 kHz sine wave, the output signal will consist of a sine wave at 5 kHz (the sum) and 1 kHz (the difference). Ring modulation seems to work best with percussive, non-pitched sounds. Typical parameters include modulation frequency, modulation depth, modulation waveform (triangle, sine, square, etc.) and unprocessed/processed mix. Note that unlike delay, tremolo, and autopan, the modulation is not subsonic but within the audio range.

Exciter

An exciter increases brightness without necessarily adding equalization. The result is a brighter, "airier" sound without the stridency that can sometimes occur by simply boosting the treble. This is often accomplished with subtle amounts of high-frequency distortion, and sometimes, by playing around with phase (a topic too esoteric to go into here). Usually there will only be one or two parameters, such as exciter mix and exciter frequency. The former determines how much "excited" sound gets added to the straight sound, and the latter determines the frequency at which the exciter effect begins.

Vocoder

A vocoder creates "talking instrument" and many other types of effects. Analog vocoders were relatively complex and costly; the versions included in some very upscale multieffects are done entirely through digital technology. Though this is not a particularly common effect, there are so many creative possibilities that it's worth discussing in some detail.

A vocoder has two inputs: one for an instrument (the *carrier* input), and one for a microphone (or other signal source; this called the *modulator* input). Talking into the microphone superimposes vocal effects on whatever is plugged into the instrument input.

Fig. 3-22 shows the block diagram for a simple vocoder. The microphone feeds four different parallel filter sections; each one covers a specific part of the audio spectrum. This part of the vocoder is electronically similar to a graphic equalizer. We need to separate the mic input into these different filter section because with human speech, different sounds are associated with different parts of the frequency spectrum.

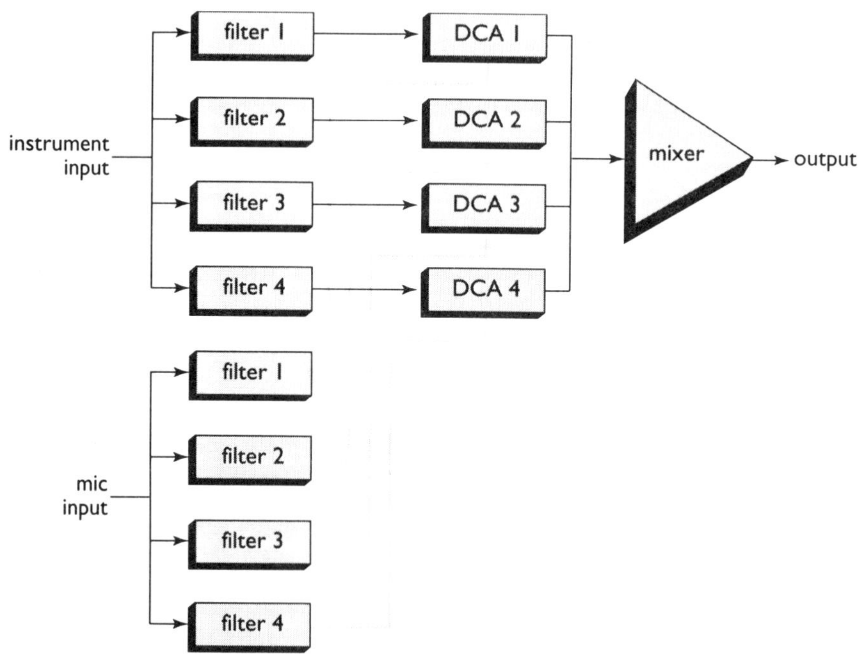

Fig. 3-22: Vocoder block diagram. Most vocoders have more than four filter stages for greater intelligibility when using voice.

For example, an "S" sound contains lots of high frequencies. So, when you speak an "S" into the microphone, the higher frequency filters fed by the mic will have an output when an "S" sound occurs, while there will be no output from the lower frequency filters. On the other hand, plosive sounds (such as "P" and "B") contain lots of low frequency energy. Speaking one of these sounds into the microphone will give an output from the low frequency filters, with little signal coming out of the higher frequency filters. Vowel sounds produce outputs at the various midrange filters.

But this is only half the picture, so let's proceed over to the instrument input and get the rest of the story.

Like the microphone channel, the instrument channel also splits into several different filters; these are tuned to the same frequencies as the filters used with the microphone input. However, these filters include digitally-controlled amplifiers (DCAs) at their outputs. The DCAs respond to the signals generated by the mic channel filters; more signal going through a particular mic channel filter raises the DCA level.

Now consider what happens when you play a note into the instrument input while speaking into the mic input. If an output occurs from the mic's lowest frequency filter, then that output controls the lowest instrument filter's DCA, and passes the corresponding frequencies from the instrument. If an output occurs from the mic's highest frequency filter, then that output controls the DCA of the highest instrument filter, and passes any instrument signals present at that frequency.

As you speak, the various mic filters produce output signals which correspond to the energies present in your voice. By controlling a set of equivalent filters connected to the instrument, you superimpose a replica of the voice's energy patterns on to the sound of the instrument plugged into the instrument input. This produces highly accurate and intelligible vocal effects. In many cases, the mic input is referred to as the "analyzed" input, since this is the signal which is "analyzed" in order to superimpose the vocal effects on to the instrument sound.

Vocoders are good for much more than talking instrument effects. For example, there's no law that says you must stick a microphone into the mic input—you could play drums into the microphone input instead of voice, and use this to control a keyboard. When you hit the snare drum, that will activate some of the midrange vocoder filters; hitting the bass drum will activate the lower vocoder filters, and hitting the cymbals will cause responses in the upper frequency vocoder filters. So, the keyboard will be accented by the drums in a highly rhythmic way. This also works well for accenting bass parts with drums.

For best results, the instrument signal should have plenty of harmonics. Otherwise, the filters won't have much to work on.

Effects Loops

This isn't really a multieffects module, but rather, two jacks on the back that let you plug an external effect into the multieffects signal chain. This is intended to accommodate some unique effect (such as a vintage wa-wa pedal) that the multieffects cannot emulate. Often the position of the effects loop can be varied within the signal chain and programmed as part of a patch.

MIDI Parameters

If your multieffects responds to MIDI, there will be a MIDI module that includes several MIDI-related parameters. The following are the most common. If any of these concepts are not familiar, be sure to read Chapter 5 on MIDI and MIDI applications.

MIDI *mode* chooses between Omni (responds to messages coming in on any channel) and Poly (responds to messages coming in on a selected channel—see next).

MIDI *channel* selects a channel from 1-16.

Program change map lets you create a table of what program should be selected in response to an incoming program change command. This is useful if you want, for example, program change 34 to call up a program other than 34. There may also be the option to turn response to program changes on and off.

Continuous controllers are handled differently from unit to unit. This setting may be part of a program, where you're presented with a list of parameters included in the program, and given the opportunity to assign them to continuous controller numbers. Or, there may be a "global" continuous controller assignment for each parameter *(i.e.,* the assignments remain fixed for all programs using those parameters). Furthermore, you may be able to vary the sensitivity and scaling, or assign a footpedal to a particular continuous controller or parameter.

Software thru or *echo* is common in units that have MIDI in and MIDI out jacks but no MIDI thru. When enabled, software thru turns the MIDI out jack into a MIDI thru.

MIDI *merging* combines the data present at the MIDI in with any data generated by the multieffects.

Save/load programs. This saves program parameters to a MIDI system exclusive storage device (see Chapter 5) for later recall. You may have the option to save one program, a bank of programs, all programs, and/or the edit buffer (a separate chunk of memory where a program "lives" while it's being edited).

Reset (Re-intialize)

The reset function usually restores all parameters to their original factory defaults. For more information, see Chapter 7 on Troubleshooting.

CHAPTER 4

Creating Your Own Algorithms

Although many multieffects arrange their available effects in fixed algorithms, some devices give you the flexibility to create your own effects algorithms. This chapter gives some guidelines on which effects should follow or precede other effects in the signal chain.

Series and Parallel Effects

There are two main ways of hooking effects together. Fig. 4-1 shows a typical *series* connection of effects, so called because the effects string together serially, one after another. The instrument plugs into effect 1's input, effect 1's output plugs into effect 2's input, effect 2's output plugs into effect 3's input, and so on.

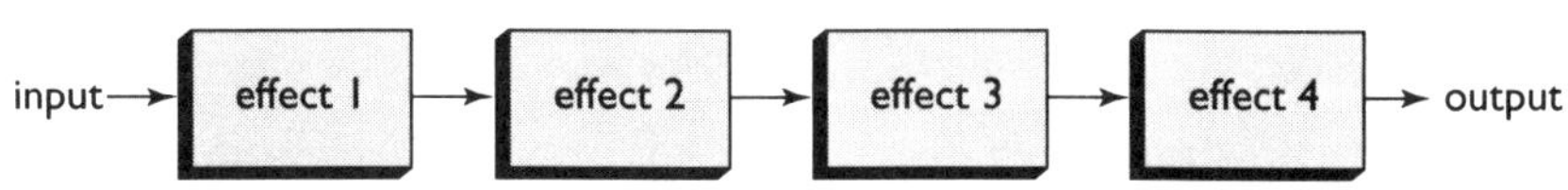

Fig. 4-1: Four effects connected in series.

Fig. 4-2 shows a *parallel* effects combination that splits the signal into the inputs of effects 1 and 2. Mixing the outputs of the two effects gives the combined (paralleled) sound of these effects. This connection is a little more complex than series connections, because it requires a mixer.

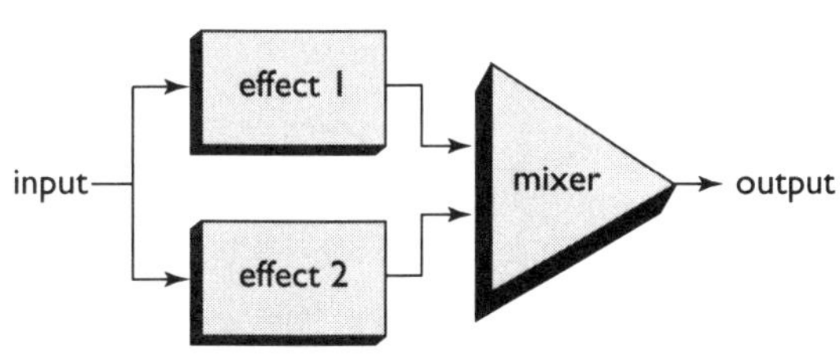

Fig. 4-2: Two effects connected in parallel and mixed into a single output.

Another variation is the *series/parallel* effects combination. Fig. 4-3 shows two series effects feeding two other groups of series effects (effects 3+4 and effects 5+6); the groups themselves connect in parallel.

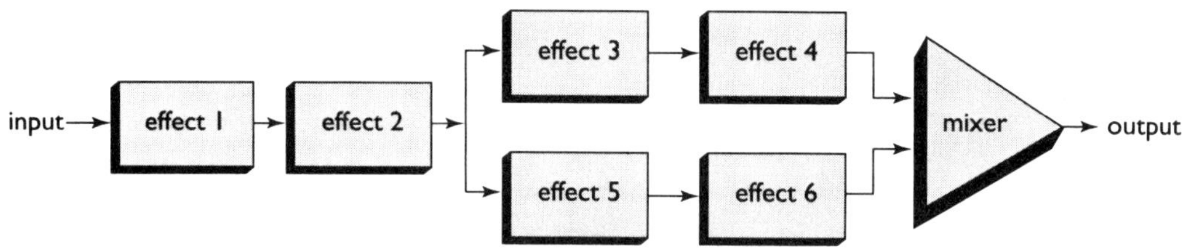

Fig. 4-3: A series/parallel combination of effects.

These configurations represent an almost unlimited way of connecting effects together for a customized sound. In fact, maybe the possibilities are just too great, because people are often confused about topics such as whether compression should go before or after distortion, which effects work best in parallel, whether a noise gate should go earlier or later in the signal chain, and so on.

Actually, there is no "right" way to create algorithms. Different combinations of different effects make different sounds, some of which you might like, some of which you might not. The right way to connect effects is purely a matter of what sounds good to you, so experiment! That's why multieffects are programmable.

Nonetheless, it's a good idea to at least establish a recommended point of departure until you feel comfortable enough with multiple effects to come up with your own algorithms. We'll start off with some general rules, then discuss series combinations of effects, and finally investigate parallel effects combinations.

Who's on First?

For many instruments, compression or distortion are generally good choices for this all-important first effect. A compressor increases sustain, which gives a "punchier" signal for subsequent stages. Also, a compressor is a unique kind of effect since if it's not overcompressed, the effect is fairly subtle and can be left in the signal chain at all times. Compression is used routinely with voice, bass, piano, and percussion.

Series Effects Combinations

When determining an order for effects, start off by considering a series effects chain as consisting of three subsections that occur in this order:

- Modules that alter dynamic range (compression) and/or synthesize frequencies (distortion, pitch transposition)
- Equalization (tone controls) to further shape the sound
- Time-altering devices (delay, chorus, reverb, etc.) to add ambience and animation

In addition, you may want to close out the chain with extra equalization to give an overall tonal tweaking, and a noise gate to cut out noise.

The order in which effects occur can make a huge difference to the overall sound. For example, distortion before equalization sounds very different compared to equalization before distortion. Here are some specific examples of what to expect when you connect two effects in series. If you are not clear about the functions of each of these effects, refer back to Chapter 3.

- **Compressor before distortion.** With guitar, this increases sustain and gives a more consistent distortion timbre.

- **Distortion before compressor** tends to be a bit noisy but provides a somewhat "gentler" sound compared to compressor before distortion.

- **Equalization before distortion.** As mentioned in Chapter 3, this can make distortion seem more "touch-sensitive" by causing boosted frequencies to distort sooner than non-boosted ones. Also, since distortion synthesizes harmonics, the overall sound may be too bright. Pulling back on the highs before feeding the distortion can help control this.

- **Distortion before equalization.** In this combination, the equalization tailors the timbre of the distorted sound. In general, having equalization both before and after distortion gives the most flexibility.

- **Pitch transposer before compressor** compresses the transposed signal along with the original signal, which often produces a more "unified" sound.

- **Pitch transposer before distortion.** Distortion generally doesn't handle polyphonic input signals very elegantly, and will tend to produce a nasty type of distortion for intervals other than octaves and fifths.

- **Distortion before pitch transposer** gives a clean harmony sound since the pitch transposer synthesizes a second distortion signal.

- **Distortion before echo** gives a clear echo sound.

- **Echo before distortion** gives a dirty echo sound since the echoes meld together when distorted.

- **Echo before noise gate.** The noise-gating action will also cut out some of the low-level echoes, which may not be desirable. Patching the noise gate before echo gives a more natural echo sound.

- **Envelope controlled filter after distortion or compression.** Because distortion and compression reduce the dynamic range, the envelope is essentially either on or off, which doesn't make for a very interesting envelope controlled filter effect. For best results, put an envelope controlled filter early in the signal chain.

- **Distortion before flanger or chorus.** Flangers and choruses give the most intense effect when they're fed a signal with lots of harmonics (overtones). Distortion delivers a harmonically rich sound that accents the flanging or chorusing effect.

- **Flanger or chorus before distortion.** The distortion pretty much masks the subtlety of the flanger or chorus effect. In fact, with extreme amounts of distortion, it may seem as if there is little, if any, flanger or chorus effect.

- **Reverb before equalization.** Boosting lower frequencies gives a larger, more "distant" reverb sound. Boosting higher frequencies gives more presence, resulting in a subjectively closer reverb sound.

- **Delay before reverb.** If your digital reverb doesn't include pre-delay, add a relatively short (30 to 80 ms) delay prior to the reverb.

- **Reverb before chorus unit.** Following reverb with light chorusing can impart a shimmering, full sound.

As an example of how to apply this knowledge, suppose you want a flexible, general purpose effects algorithm for power-chord rhythm guitar. Here's one possibility:

compressor » equalizer » distortion » equalizer » chorus » noise gate » reverb

The compressor adds sustain, and the EQ/distortion/EQ combination allows for a very precise tailoring of the distorted sound. The chorus helps diffuse the overall sound somewhat; the noise gate cuts out any low level noise created by the distortion, and the reverb "smoothes over" choppiness contributed by the noise gate.

To change this to a rock lead guitar algorithm, adding pitch transposition (to create harmonies) and delay can be useful. The chorus, since it diffuses the sound, is probably unnecessary, particularly if there are memory limitations and it's necessary to choose between chorus and delay. Similarly, if memory limitations mean you can't have reverb and pitch transposition simultaneously, I'd recommend eliminating the reverb.

compressor » equalizer » distortion » equalizer » pitch transposition » noise gate » delay » reverb

Parallel Effects

Parallel effects combinations can provide a greater degree of subtlety than series combinations. As one example, putting bass through a chorus and then a wa-wa will give a thin sound because the wa-wa removes the bass "bottom." Placing the wa-wa in parallel with the chorused signal *adds* the filtered effect to the chorused bass sound (which doesn't remove the low end).

Parallel effects chains are also a good way to create a stereo image, since one leg of the chain can provide one channel and the other leg, the other channel. For example, suppose you feed a piano into two 8-band graphic equalizer modules connected in parallel. If you set bands 1, 3, 5, and 7 to maximum and bands 2, 4, 6, and 8 to minimum in one channel, and do the reverse for the other channel (bands 1, 3, 5, and 7 to minimum with bands 2, 4, 6, and 8 to maximum), you'll create a cool-sounding stereo spread.

Here's a very dramatic example of the advantage of using two effects in parallel. This technique uses two paralleled delay lines to give drastically improved flanging sounds compared to using a single delay.

As mentioned in the previous chapter, flanging mixes two signals, one straight and one with variable delay. As the time delay difference between the two signals gets smaller and smaller, you hear the characteristic "jet airplane taking off" sound of flanging until the time delay difference reaches 0 seconds. This creates the "peak" of the flanging effect, and is called the "through-zero" point.

Unlike tape-based flanging, digital delays can't sweep down to 0 seconds; there will always be a minimum delay time for a signal going through the delay line, so the flanging effect will never reach the through-zero point, and never give the "peak" that tape-based flanging can.

However, there is a workaround using parallel effects. Fig. 4-4 shows the basic setup for through-zero flanging. Set Delay 1 to a small, fixed delay (*e.g.*, 2 ms). Use a triangle wave to sweep Delay 2's time from 2 ms (Delay 1's delay time) to around 20 ms or so.

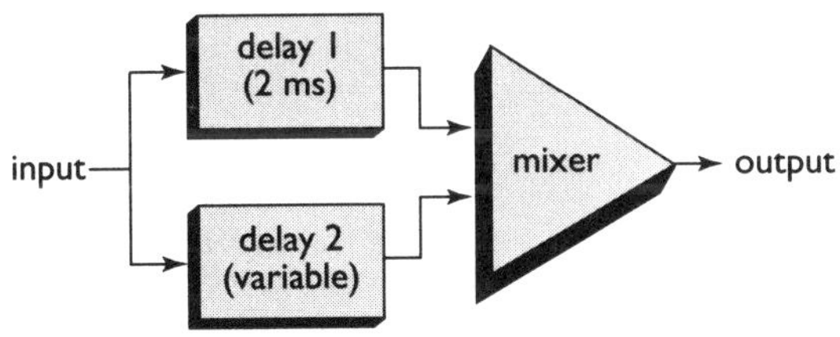

Fig. 4-4: Basic algorithm for through-zero flanging effects.

We've now satisfied the requirements for through-zero flanging since as Delay 2 sweeps, the time delay difference between it and Delay 1 gets smaller and smaller until it reaches 0 seconds. Because Delay 1 is set to such a short delay, playing through it sounds like it's happening in real time, even though there is a slight delay.

Note that both delays should be set to the same levels, and to *delayed signal only.* Remember, Delay 1 provides the "straight" signal.

CHAPTER 5

Multieffects and the MIDI Connection

Most modern multieffects include MIDI (Musical Instrument Digital Interface) connections, which (as you'll find out) is a Good Thing. Many musicians think that MIDI is just for keyboard players; years ago, that was more or less the case. Now, though, MIDI is much more universal. Electric and acoustic musicians (from speed-thrash-metal-postpunk guitarists to ballad singers) as well as recording engineers can use MIDI to control multieffects, and in the process, sculpt a unique and fresh sound.

Whether live or in the studio, MIDI-controlled signal processing represents a new frontier that has been little explored. In particular, guitar players—whose sonic arsenal has been augmented by MIDI multieffects, program change footswitches, and control pedals—can benefit greatly from recent advances in signal processing technology.

The complete MIDI spec is fairly "deep" and complex. Fortunately for MIDIphobes, there are only a few parts of the spec that relate to multieffects control, so there's not really that much to learn. Let's start with MIDI basics.

MIDI Basics

A MIDI system requires something that sends MIDI messages (such as a footpedal or footswitch that says "make the instrument louder," "select this effect," "add more echo," etc.) and something that receives these messages and acts on them, like a multieffects (Fig. 5-1) or MIDI-controlled amplifier.

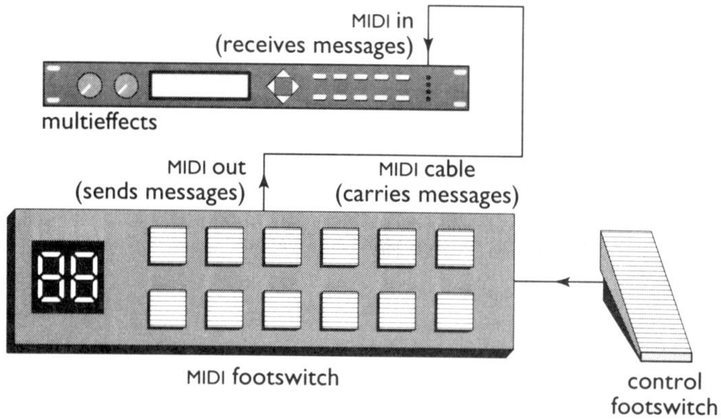

Fig. 5-1: A basic MIDI system. Circuitry inside the footswitch converts the pedal motion to MIDI data; this is sent along with the footswitch data to the multieffects. In the studio, sequencer control usually replaces pedals and footswitches.

Don't let the term "Musical Instrument Digital Interface" throw you; just remember that the key word is "interface." MIDI's main purpose is to allow musical machines to communicate musical data to each other, such as which note is being played, which pedal is being moved, and so on. Stripped to its basics, you can think of MIDI as a catch-all name for the process of sending control messages from one device *(i.e.,* a footswitch or sequencer) to another device *(i.e.,* a multieffects) over a MIDI cable.

There are many different kinds of MIDI messages, most of which relate to keyboards, sequencers, drum machines, lighting controllers, tape recorders, and other gear we can ignore. If you're controlling multieffects, 99% of the time you need to know about only two kinds of MIDI messages: *program changes* (which call up different multieffects programs) and *continuous controller* messages (which alter effects parameters in real time).

Some people think MIDI involves bits and bytes and techno talk like that. But you don't have to know about microwaves to use a microwave oven, nor do you have to know about superheterodyne circuitry to listen to the radio, and you don't need to know bits and bytes to use MIDI. Of course, if you're a naturally curious type and want to investigate the subject further, be my guest (in particular, check out the Amsco publications *MIDI for Musicians, MIDI for the Professional,* and *Digital Project for Musicians).* These days, however, MIDI equipment has reached a degree of sophistication that, paradoxically, makes it easier than ever to use.

MIDI Connections

Virtually all MIDI-equipped signal processors have a *MIDI in* and *MIDI out* jack. There may also be a *MIDI thru* jack, which provides a duplicate of the signal at the MIDI in jack. Thru jacks are not required by the MIDI specification; we'll cover their use later. Fig. 5-2 summarizes what these jacks do.

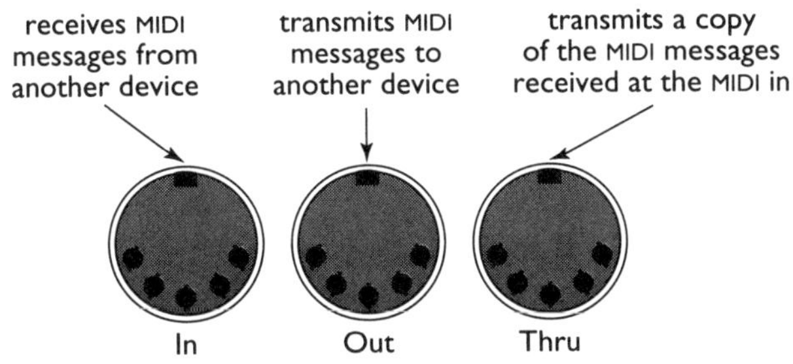

Fig. 5-2: Typical MIDI connections found on a multieffects' rear panel.

Channels and Modes

Most MIDI messages are "stamped" with a unique channel identification from 1 to 16. This allows a single cable to carry messages for up to 16 different devices; transmitters can send data over a particular channel, and receivers can tune in to a specific channel.

There are also two MIDI modes used by most multieffects that determine their response to channelized data.

- *Omni* mode accepts data coming in over *any* channel. In other words, regardless of the channel ID, the multieffects will attempt to act on any incoming data.

- *Poly* mode is the most commonly used multieffects mode. A receiver in poly mode will be set to one of the 16 MIDI channels and receive only messages intended for that channel. Thus, two MIDI receivers set to receive different channels could be monitoring the same data stream, but be controlled independently of each other.

Generally, a multieffects will default to one of the following combinations of modes and channels:

- Poly mode, with the channel set to whatever channel was selected last (in other words, the device remembers any channel assignment you made, even when power is turned off). This is probably the most common option.

- Omni mode, so that the multieffects will respond to anything that provides MIDI data.

- Poly mode, with a default channel setting (such as channel 1).

The bottom line: if you have a MIDI transmitter (like a footswitch) controlling a multieffects, just make sure both units are set to the same channel, or the multieffects is set to omni, and the transmitter and receiver will be able to communicate.

About Program Changes

To understand program changes, take a trip with me down memory lane to the mid-70s, when disco ruled, synthesizers were not yet programmable, and guitar effects were starting to progress beyond funny little boxes that ate batteries and burped noise. Guitarists discovered early on that the hippest control on any signal processor was the in/out footswitch because it allowed you to bring an effect in as needed (distortion is wonderful, but not all the time).

As more boxes were introduced—compressors, phase shifters, echo units, wa-was, etc.—musicians started connecting all these boxes together with patch cords. And that's where the trouble began: if you wanted to bring in multiple effects simultaneously, you had to hit a bunch of footswitches at once (Fig. 5-3). This may not be a problem for Bigfoot, but for the rest of us, fancy tap dancing became the order of the day.

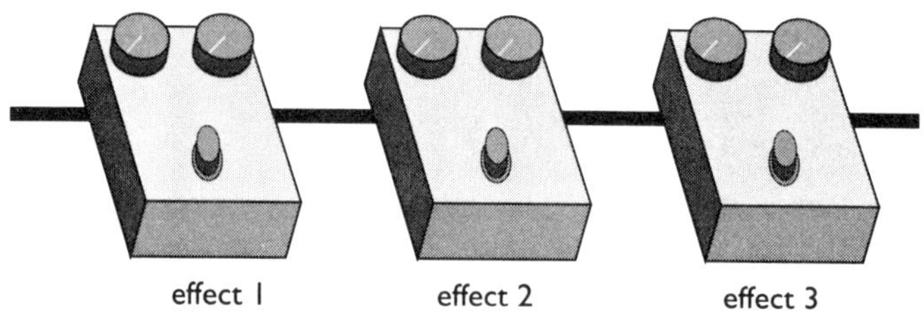

Fig. 5-3: Multiple effects systems used to mean lots of little boxes strung together with patch cords.

Manufacturers responded with a variety of methods to make life easier on stage (as well as separate us from our disposable income). One of these involved building several signal processors into one box, along with their associated footswitches. A more advanced approach was to build all the signal processor electronics into a box, and run a cable to a remote footswitch assembly with multiple footswitches (Fig. 5-4). Further along the evolutionary path, sometimes these footswitches would include a master bypass button that would let you set up a particular combination of effects, then bring the whole mess in and out by pressing a single button.

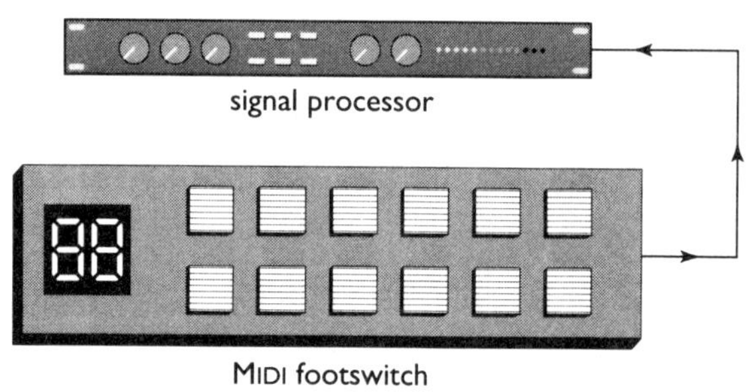

Fig. 5-4: Separating the footswitch and the signal processing circuitry allowed for greater on-stage convenience.

Unfortunately, these footswitches often controlled actual audio lines, which meant that the cable going from the signal processor to the footswitch assembly carried your delicate audio signal, making it susceptible to picking up hum and other noises. Some manufacturers got around this by building electronic switching into the main effects box and just sending control signals to these electronic switches. But that solution was far from ideal, because each manufacturer had a different way of implementing the footswitch control circuitry. Therefore, a Brand A footswitch could work only with Brand A effects, and a Brand Z footswitch could work only with Brand Z effects. If you wanted to trade up to a newer effects box, you had to junk the footswitch and get a new one.

Program Changes to the Rescue

Then came the introduction of cheap computing power. This meant that smaller businesses, not just giant institutions, could make incredible mistakes and blame them on computers. But it also meant that signal processors could use computer power to manage your effects in a much more intelligent way, and MIDI was the ticket.

Rack mount multieffects started to appear where you could dial up the sounds you wanted, using particular combinations of effects and control settings, and save all these settings as programs. No longer did you have to twist dials and push multiple footswitches to go from one sound to the other—select a program, and the sound would be called up exactly as you programmed it.

When the MIDI spec was drawn up, provisions were made for 128 MIDI program change messages. This is why many signal processors offer 128 programs. As usual, people wanted more—so now an addition to the MIDI spec, Bank Select messages, can select up to 16,384 banks of 128 programs for a grand total of over 2,000,000 programs (that should hold you for a while).

MIDI Program Change Footswitches

Now let's consider a modern MIDI-based footswitch setup, as shown previously in Fig. 5-1. The multieffects' MIDI in jack receives MIDI messages; the footswitch's MIDI out jack transmits MIDI messages.

Pressing on an individual footswitch transmits a MIDI program change, and the signal processor calls up the program associated with this program change number. If pressing the footswitch sends out program change #43, the signal processor selects program #43. It's simple, and what's more, it's standardized—you can use a Brand A footswitch with a Brand Z signal processor because all manufacturers have agreed to use the same message format. When *any* footswitch sends out program change #43, it will look the same as far as any signal processor (receiver) is concerned. (However, there is one complication; see sidebar.)

Invasion of the Strange Program Change Numberings

MIDI allows for 128 program changes, but doesn't specify how these are to be numbered. Some manufacturers might number programs as 1-128, 0-127, or as banks (*e.g.*, eight banks of 16 programs).

Suppose you program a MIDI footswitch to send out program change 15. Units that number programs from 0-127 will interpret this as 15, but if the program numbering scheme is 1-128, this will show up at the multieffects as program 16. Or you might even get something bizarre like Program 2-7 (bank 2, program 7) if the device has several banks of eight programs each.

It's worth running a test to see how your various effects respond, and making up a conversion chart that shows how program changes generated by your MIDI transmitter are interpreted by your various MIDI receivers.

There are many ways to implement how a footswitch sends out program changes. The least practical (and certainly the most costly) way would be to have 128 buttons, each of which sends out a single program change command. One possible alternative would be to have eight footswitches, and two up/down footswitches that step through 16 banks. Bank 1 could have the eight footswitches send out program changes 1-8, Bank 2 program changes 9-16, Bank 3 program changes 17-24, etc.

An even more versatile approach lets you program each bank's switch to send out whatever program change you want. For example, suppose you use signal processor presets 12, 15, 37, 24, and 6 in a tune. You could program one footswitch bank so that footswitch 1 sends out program change 12, footswitch 2 transmits program change 15, footswitch 3 sends out program change 37, and so on. You can then hit each footswitch in a convenient, logical order rather than jump between different banks and hit different footswitches.

MIDI Program Change "Mapping"

A one-to-one correspondence between footswitch messages and multieffects program numbers is not always desirable. For example, you may want to call up the same program in several different footswitch banks, but copying the program to several different locations in the multieffects wastes program memory.

One way to set up a different correspondence between footswitch messages and multieffects programs is called *mapping,* which involves creating a *program change table* in the multieffects that specifies which program will be called up in response to a particular program change number. For example, you might program a table where program change message 1 calls up signal processor program 16, program change message 2 calls up signal processor program 51, program change message 3 calls up signal processor program 47, and so on. Fig. 5-5 shows an example of a program change table.

MIDI transmitter sends:	this calls up program:
1	16
2	51
3	47
4	17
5	62
6	43
7	18
etc.	

Fig. 5-5: Many multieffects let you create a program change table that translates incoming program changes to a different program number.

Mapping is also useful with devices that have more than 128 programs but do not allow for Bank Select. You could map, say, program 125 to program 265. Although this would still only allow you to access 128 programs, they could come from a much larger "pool" of available programs.

Daisy-Chaining Multiple Units with MIDI Thru

Many MIDI devices include a MIDI thru jack that transmits a copy of the data appearing at the MIDI in. (Another option is that the MIDI out might function as a MIDI thru in normal use, but be used as a MIDI out for special functions such as system exclusive dumps, as described at the end of this chapter.) A single MIDI transmitter can therefore address multiple units by "daisy-chaining" one unit's MIDI thru connection to the next unit's MIDI in (Fig. 5-6).

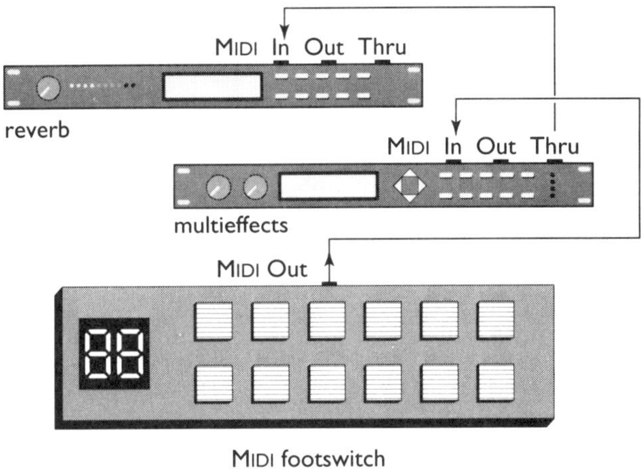

Fig. 5-6: "Daisy-chaining" makes it easy for a single MIDI transmitter to control several MIDI receivers.

Incidentally, daisy-chaining more than four or five devices together is not a good idea because each time the MIDI signal goes through the MIDI thru connection, there can be a slight degradation of the signal. By the time messages reach the end of the chain, there could be enough "distortion" to create erroneous data. If you have lots of effects, it's good practice to use a separate MIDI thru box that can take a MIDI output and split it among several devices (Fig. 5-7).

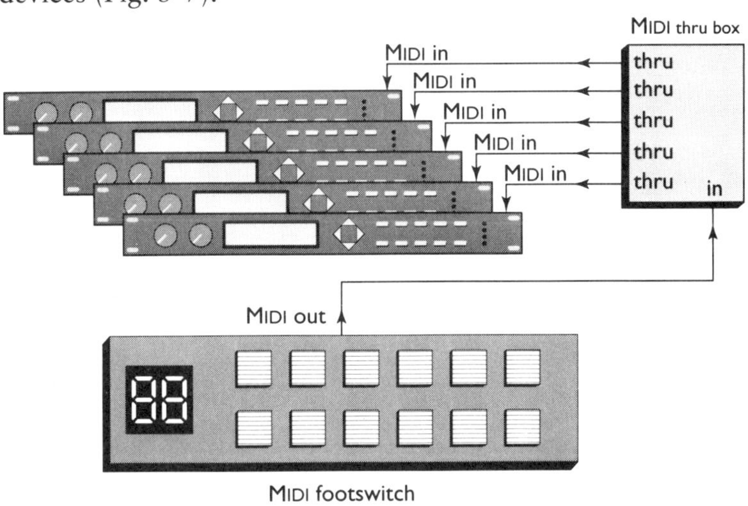

Fig. 5-7: Using a MIDI thru box to drive multiple units and minimize data distortion.

Managing Program Changes With More Than One Multieffects

When using multiple devices, a complication occurs if you obtain your favorite sounds by combining different programs on different units *(e.g.*, program 16 from multieffects device A, program 12 from device B, and program 120 from device C). If the footswitch (or other MIDI transmitter) sends out only a single program change command, then you're not going to be able to call up the three different sounds you want at the same time. Here are four possible options:

- Copy the programs you want to use simultaneously into identically-numbered program locations. In the example above, if you copied all desired programs to program 1 on their respective multieffects, then calling up program 1 would select the appropriate effect on all three units.

- Use the re-mapping option mentioned earlier to map a single program change number to the desired target programs.

- Program each unit to poly mode and to receive on its own MIDI channel, then send individual program changes over individual channels. This means that your footswitch must be able to send multiple commands from one footswitch (in this case, different program changes over different channels).

- Use a hardware MIDI data processor to convert a single incoming program change received over one channel into multiple program changes on multiple channels.

Deciding which method to use depends on your hardware setup. Suppose a MIDI footswitch can send out data on only one channel at a time; the first three methods are your best options. A footswitch that can send multiple program changes over multiple channels is a good candidate for method 3, which is probably easier than remapping programs or copying them to different locations. Although this means you won't need the external box mentioned in method 4, you will need to spend some time programming your footswitch to send out the correct data on the correct channels.

About Continuous Controllers

Changing from one program to another is a good start, but sometimes you'd like to vary a particular parameter within an individual program (delay feedback, filter frequency, distortion drive, etc.). A wa-wa sound is a good example of changing a particular parameter (filter frequency) in real time.

When you call up a program's parameter, you usually change its value by turning a front panel knob, or perhaps pushing buttons to enter a particular value for that parameter. If you're using a knob, one obvious way to provide pedal control would be to remove the knob from the front panel, put it in a pedal, and run a cord between the pedal and the unit.

Unfortunately, that doesn't let you change multiple parameters at the same time, and involves some hardware hacking. Fortunately, most signal processors will let you control parameters remotely by sending them MIDI continuous controller commands over a MIDI cable.

The idea of *continuous controllers* came about because synthesizers have pedals, knobs, levers, and other physical "controllers" that alter some aspect of a synth's sound over a continuous range of values (this is why they're called *continuous controllers,* as opposed to a controller such as an on-off switch, which only selects between two possible values). Non-keyboard musicians can use other controllers, such as footpedals or data entered into a sequencer, to alter some aspect of a signal processor's sound.

Unlike a program change, which is a single event, continuous controllers generate a series of events, such as a volume fadein (each event raises the volume a bit more than the previous event), or change in some other parameter *(e.g.,* increasing chorus depth, or altering the wa-wa filter frequency).

Like program changes, continuous controller messages are transmitted over a MIDI output and received by a MIDI input. The transmitter usually digitizes the physical controller motion into 128 discrete values (0-127). As one example, consider a footpedal that generates continuous controller messages. Pulling the pedal all the way back typically generates a value of 0. Pushing down on the pedal increases the value until at midpoint, the pedal generates a value of 64. Continuing to push on the pedal until it's all the way down generates a value of 127.

Note that continuous controller transmitters only send messages reflecting a change; for example, leaving a pedal in one position doesn't transmit any messages until you change the pedal's physical position.

Continous Controller Numbers

MIDI "tags" each continuous controller message with an ID from 0 to 127. Don't confuse this with channel IDs; each channel can support up to 128 controllers, so something like a controller 7 message appearing over channel 2 is independent from a controller 7 message appearing over channel 3. Therefore, a signal processor with 100 different parameters could have each assigned to a unique controller number (1 for reverb first reflection time, 2 for reverb high-frequency decay, 3 for reverb low frequency decay, etc.).

When controlling a signal processor via continuous controllers, the basic idea is to assign a particular signal processor parameter (echo mix, chorus speed, distortion drive, etc.) to a particular continuous controller number. Then all you need is a continuous controller transmitter (such as a pedal or MIDI sequencer) that generates controller messages of the same number.

At the receiving end, the parameter being controlled changes in response to incoming message values. For example, if you're controlling delay feedback and it receives a value of 0, feedback is at minimum. When it receives a value of 64, the feedback might be halfway up and upon receiving a value of 127, the feedback might be up all the way. The reason for saying "might be" is that some signal processors let you *scale* and/or *invert* the values, as explained later.

Here are some other examples. If a device's level parameter is assigned to controller 7, and a footpedal can generate controller 7, hook the pedal's MIDI out to the device's MIDI in and your pedal will control level (providing that both are set to the same channel, of course). If you assign the chorus depth parameter to controller 12 then set the pedal to generate controller 12 data, the pedal will vary chorus depth.

Controller Assignment Protocols

There are two common ways in which signal processors assign controller numbers to parameters:

- **Per Program.** Each program has an assignable parameter (or group of parameters) that can be controlled by continuous controller messages. For example, controller 5 might vary chorus depth in one program, or echo feedback in another. This is useful if you have a footpedal that generates a specific continuous controller number—say, controller 1—that is hard to change. Suppose you want to control compressor level in program 1, echo time in program 2, and filter frequency in program 3. Assign the desired parameter in each program to controller 1, and when you call up the program, the footpedal will control the assigned parameter (compressor level in program 1, echo time in program 2, etc.).

- **Global.** With "global" controller assignments, each multieffects parameter (or at least most parameters) has a continuous controller number that is fixed, either at the factory, or according to whatever assignment you give it. For instance, if echo feedback is assigned to controller 12, then every patch that uses echo feedback will have that parameter respond to controller 12 messages. This makes it necessary to select the desired controller number at the transmitter to accommodate the receiver, whereas with the per-program approach, it is more common to change the target controller number at the receiver.

With many multieffects, you can assign several parameters to the same controller number so that, for example, a single pedal motion could increase the level *and* reverberation time *and* boost the upper midrange.

Other Control Message Options

In addition to responding to continuous controllers, some multieffects respond to other MIDI control messages. None of these has a controller number, as each is deemed important enough to be its own distinct class of message.

- **Pitch bend.** Most synthesizers have some type of modulation wheel or lever that allows for bending note pitch (like bending a string).

- **Velocity.** This indicates the dynamics of a playing a keyboard by measuring how long it takes for a key to go from full up to full down. The assumption is that the longer it takes for a key to go down, the more softly it's being played.

- **Pressure.** After a keyboard key is down, pressing on the key produces an additional pressure message. A common application is to use this to introduce vibrato to a sound that's being sustained, or to change a sound's tone *(e.g.,* make it brighter).

- MIDI **timing messages.** MIDI also includes timing-related messages for synchronization among rhythmically-oriented devices such as drum machines and sequencers. This is not something that's important for multieffects, with one exception: some delay modules can set their delay time according to MIDI clock messages.

  A MIDI clock message occurs 24 times per quarter note, and provides the master metronome signal to which all devices synchronize. Multieffects that sync to MIDI clocks do so by counting the number of clock pulses and adjusting the delay time accordingly. For example, if you want a quarter note delay, the multieffects will set its delay time to equal 24 MIDI clock pulses. For a half-note delay, it would count 48 clock pulses. A sixteenth-note delay would be 6 MIDI clock pulses, and so on.

  All of this is pretty much "transparent" to the user. Simply plug the MIDI out from a drum machine, sequencer, or other rhythmically-oriented device into the multieffects MIDI in, and set the delay time to a specific note value. The multieffects' computer will take care of the rest.

How Signal Processors Handle Controllers

There are several ways that differing units decide how continuous controllers will "take over" from a pre-programmed setting. One approach is to add to (or subtract from) the programmed setting; usually *scaling* and *inversion* parameters will be available.

Scaling determines how far the parameter can vary from the programmed setting in response to a given amount of controller change. Inversion sets whether increasing controller values will increase (+ scaling) or decrease (-scaling) the parameter value. Often these are combined into one number, such as +50 (which represents 50% scaling of full value in a positive direction), -37 (37% scaling of full value in a negative direction), etc.

Fig. 5-8 shows an example of an input control signal scaled to +100, +50, and -100. Each one covers a different range of the available parameter values. Greater controller amplitudes increase or decrease the programmed parameter value, depending on whether the polarity is positive (+) or inverted (-).

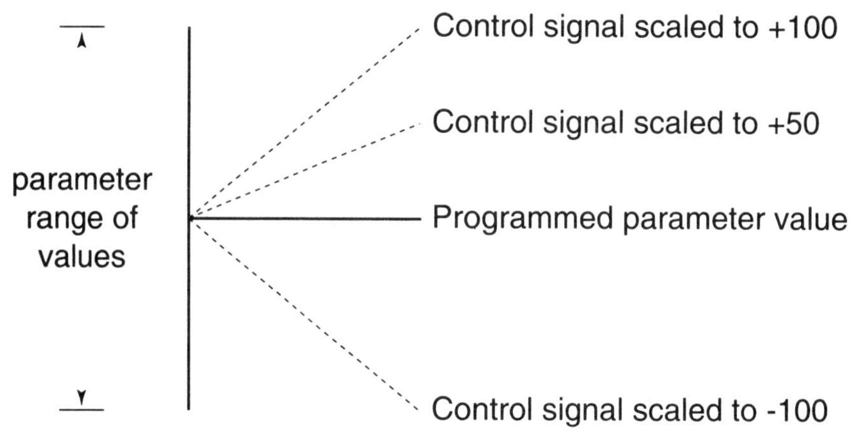

Fig. 5-8: Controller scaling and inversion affects the degree to which the receiver responds to continuous controller messages.

If you encounter a situation where the continuous controller messages don't change a parameter very much, check the programmed value of the parameter being controlled. If its value is close to minimum or maximum, there may not be enough "headroom" for the controller to make much difference, especially with scaling close to 100%.

Another design approach is to simply have the parameter follow the incoming controller value. An incoming controller value of 0 would set the parameter to its minimum value, and 127 to its maximum. This works well with sequencers, where it's usually easy to adjust controller values. With footpedals, this approach is less applicable since you might not always want the footpedal to cover a parameter's maximum possible range. For example, if you're using a pedal to vary equalization, always going from maximum boost to maximum cut may not be as useful as covering a narrower range.

Yet another method will not let parameters respond to continuous controllers until the controller passes through the preprogrammed value, at which point the parameter follows the controller messages. This is helpful when switching between programs where a footpedal is programmed to control different parameters. The parameter will stay as originally programmed until you start using the pedal and go past the existing setting.

Using Controller Changes Instead of Patch Changes

Sometimes a patch change is too abrupt to give the desired effect. For example, on one tune I use the same basic patch throughout but change the distortion drive for rhythm and lead parts. During a lead section, continuous controllers also determine the amount of echo feedback, echo mix, and sustain. All of these changes happen over time *(e.g.,* as the solo progresses, the amount of echo feedback increases) which sounds very natural. Incidentally, these changes can all be triggered by a sequencer (see the next chapter) so there's no pedaling or footswitching required.

Summary: Assigning Continuous Controllers in a Nutshell

To recap, here's how to assign multieffects controllers. We'll use a MIDI footpedal as an example of a typical transmitter, but the same concepts apply to using a sequencer or a keyboard's continuous controllers.

1. Set the pedal and multieffects to the same MIDI channel (1-16), or set the multieffects to omni mode (which accepts data from any MIDI channel).

2. Assign the multieffects parameter you want to control to the desired controller number. With per-program MIDI controller implementations, assign the parameter to be controlled within a program to a specific continuous controller number. With global MIDI controller implementations, find out which controller number controls a particular parameter *(e.g.,* controller 17 may control echo time, controller 18 echo mix, etc.). Read the manual for this info, and whether you can assign the controller number or whether it's fixed.

3. Program the pedal to the controller number that matches what you programmed in the previous step. Depending on your gear, the pedal may generate a fixed controller number, in which case you'll need to program the multieffects parameter to match it; or the multieffects parameter may respond to a fixed controller number, which requires programming the pedal to the same number. Or, both the multieffects and pedal may be adjustable (the most flexible option of the lot, and these days, the most common).

4. At the multieffects, adjust the controller amplitude (how much the sound is affected when you move the pedal) and scaling (whether pushing the pedal forward increases or decreases the parameter value).

5. Move the pedal, control the parameter, and have a blast!

Your Friend, the MIDI Implementation Chart

The MIDI spec suggests that any MIDI-controlled device include a MIDI Implementation Chart, which provides data about a piece of gear's MIDI performance. This will show whether a unit responds to program changes, which continuous controller commands are recognized, whether the device can save and load programs via system exclusive information, etc.

Fig. 5-9 shows a typical MIDI implementation chart for a generic signal processor. Note that O indicates a feature is implemented, and X means a feature is not implemented. From this we can see that the device:

- Receives on all 16 channels, and remembers the channel you set (even if power is turned off).
- Responds in either omni or poly mode, and transmits in poly mode.
- Does not respond to note number, velocity, pressure, or pitch bend information, but recognizes continuous controllers 1-120. The "remarks" column indicates that these are used for real time parameter control.
- Has 128 programs, numbered as 0-127.
- Transmits and receives system exclusive information.
- Recognizes MIDI clock signals, and uses these to synchronize delay times.

Model: MegaMFX
Version: 1.4

MIDI Implementation Chart

Date: 1/95

Function		Transmitted	Recognized	Remarks
Basic Channel	Default Channel	1-16	1-16	memorized
Mode Messages	Default	3	1,3	
		X	X	
		X	X	
	Altered			
Note Number	True Voice	X	X	
Velocity	Note ON	X	X	
	Note OFF	X	X	
Aftertouch	Key	X	X	
	Channel	X	X	
Pitch Bender		X	X	
Control Change		X	0-120	Used for real time MIDI control
Program Change	True #	0-127	0-127	
System Exclusive		O	O	
System Common	SPP	X	X	
	Song Select	X	X	
	Tune Request	X	X	
System Realtime	Clock	X	O	Used for delay sync
	Commands	X	X	
Aux	Local ON/OFF	X	X	
	All Notes Off	X	X	
	Active Sensing	X	X	
	Reset	X	X	

Mode 1: Omni On, Poly   Mode 2: Omni On, Mono   O:Yes
Mode 3: Omni Off, Poly   Mode 4: Omni Off, Mono   X:No

Fig. 5-9: Typical MIDI implementation chart.

Saving and Loading Programs With Sys Ex

MIDI can do a lot more than control parameters and change programs; saving and loading data via MIDI is one of the most important other applications. So, let's see how to get saved—not the religious experience, but the data inside your multieffects signal processor.

You probably don't think too much about saving data, since if you use a multieffects, it has probably been very reliable: when you turn it on, you're used to seeing all your favorite programs, including your custom edits, in their appropriate memory slots. Unfortunately, it's possible that the next time you turn on your multieffects, those programs will be gone. If you saved your data, no problem—reload it back into the multieffects and you're ready to rock. Otherwise, you may have to kiss goodbye to days, weeks, or months of parameter tweaking. (If you only use factory presets, though, they can probably be recalled. But why be normal? Program your own sounds!)

When Bad Things Happen to Good Machines

The main causes of memory loss are:

- **Dead memory backup battery.** Backup batteries have a long rated life, but that's an estimate, not a guarantee. At some point a backup battery will fail, and with it, the memory that it keeps alive.

- **Physical damage.** Dropping a unit could disconnect a wire from the backup battery or damage a component, wiping out memory.

- **RAM chip failure.** If the memory chip fails, your programs are history.

- **Human error.** It's possible to accidentally overwrite a program, or do some other Dumb Human Thing that erases valuable data.

- **Software lock up.** Sometimes, due to static electricity discharges, software glitches, or other cruel twists of fate, a device will "lock up" and the controls will have little or no effect. Re-initialization (a generally user-accessible service procedure; see Chapter 7) will often solve the problem but at the expense of wiping the RAM clean.

- **Software updates.** If you've upgraded your unit with the latest software chip, re-initialization will probably be required.

Let's Back Up a Bit

The best insurance against memory loss is backing up your data, and preferably making more than one set of backups with the second set stored in a separate location.

There are several common data backup methods (cassette interface, saving to an onboard disk drive, copying programs to a RAM card or second unit), but the most common way to back up a multieffects is by storing its data as MIDI messages in a suitable device. This method offers the advantages of reliability and—assuming you have access to one of the many kinds of sys ex storage devices—low cost.

Practicing Safe Sys Ex: How it Works

The parameters in a multieffects are all just numbers being crunched by a computer, and a program is simply a collection of parameter numbers. So, everything that needs to be saved in a multieffects is basically computer data.

The MIDI specification includes a way for manufacturers to send equipment-specific data in a format that only equipment "trained" to recognize that data will accept. This type of data is called *system exclusive* (as opposed to data that's common to a wide variety of gear) and called sys ex for short. Furthermore, this sys ex data can be sent along a MIDI cable since it follows the same basic MIDI protocols with respect to transmission rate, format, etc.

Although sys ex data is used in many ways, one of the most common applications is to translate patch parameters into sys ex. If you send this data over the MIDI out connector, and the data enters the MIDI in of a device that stores sys ex, your program data will be saved.

Sys Ex Storage Devices

There are several ways to store sys ex data:

- **Dedicated devices.** These are boxes whose sole purpose in life is to store sys ex data easily and efficiently. For example, the Alesis DataDisk and Peavey MIDI Streamer are rack units that can save sys ex data to standard disks loaded into their built-in 3.5" disk drives.

- **Keyboards.** Several keyboards let you load sys ex data into memory, then save it to the unit's internal disk drive. However, keyboards usually have some storage limitations, such as not being able to handle real large files. Also, most disk formats are incompatible—you can't take sys ex data stored on the disk drive in one manufacturer's keyboard and play it back from a different brand of keyboard. Nonetheless, if there's a keyboard player in your band, that may be all you need to back up your programs.

- **Computer software.** Many sequencer programs let you store sys ex messages as part of a sequence. Also, universal "librarian" software can store sys ex messages from a variety of gear.

Basic Procedures

Most of the time, the way to save sys ex is pretty simple:

1. Hook the MIDI out of the device whose data is to be saved to the sys ex storage device's MIDI in.

2. Set up the storage device to receive MIDI data. If there's a display, it will usually indicate that the unit awaits MIDI data. You need not worry about setting devices to the same MIDI channel, since sys ex data is channel-independent.

3. Consult your manual on how to send a sys ex data dump from your multieffects. Usually this function is somewhere in the unit's MIDI menu.

4. Initiate the data dump. Often the transmitter will show that it is transmitting data, and the receiver will confirm that it is receiving this data. The total transfer time depends on the data file's length, but will probably be well under 30 seconds.

5. If necessary, save the received data to disk or some other non-volatile storage medium.

Potential Problems

If you try to save a file that is larger than the capacity of the data storage device, the file will not be properly saved. The receiver may or may not indicate that a problem has occurred. Some effects will let you save several smaller banks of files instead of (or in addition to) one large file to accommodate storage devices with limited memory capacities.

Another problem (which occurs mostly with older gear) is that it may not be possible to initiate a sys ex dump from the unit itself. Instead, the storage device will need to "request" a dump, and you have to tell it how to do this. The multieffect's manual may contain this information, but probably not in a way that's understandable to the MIDI novice.

Most of the time, though, backup is straightforward, so there's no excuse not to save for a rainy day. Backing up data is cheap and effective insurance against losing a lot of hard work.

CHAPTER 6

Multieffects Applications

Total Effects Control in the Studio

MIDI program changes and continuous controllers can automate multieffects parameter changes in sync to a recording. This leaves you free to concentrate on your playing (or mixing) since you don't have to hit any footswitches or move any pedals. You can even apply MIDI-controlled effects to acoustic instruments recorded on tape.

A device called a MIDI *sequencer* provides this automation, so let's first talk about sequencing before describing how to apply this technique in the studio.

Sequencing Basics

A sequencer records MIDI data in computer-type memory for later playback. The process is similar to tape recording: press *record* and feed MIDI data into the sequencer; when you're done, press *play* to play the data back.

Sequencers are available as stand-alone, self-contained hardware devices, or as software you can load into a personal computer such as the PC or Mac. The computer may require an additional hardware accessory, such as a MIDI interface or sound card with MIDI capabilities, to let it communicate with MIDI devices. In addition, many keyboard "workstations" include sequencers that can record continuous controllers and program changes along with other MIDI data.

Almost all sequencers, when combined with the proper sync-to-tape hardware, can synchronize to a synchronization track recorded on analog or digital audio tape. As you play the tape, the sync track feeds information into the sequencer that allows it to remain in sync with the tape (even if it speeds up or slows down slightly, as happens with analog tape).

To understand synchronization, think of a how a band works: the drummer keeps the time, and by listening to the drummer, the other musicians remain in sync with the song. The sync track is the "drummer" to which the sequencer listens. If you enter data into the sequencer while the sequencer is synched to tape, when the tape plays back the MIDI data will play back at the same point you entered it.

Debunking a Myth

A common misconception is that even in the studio, to automate your effects you'll have to play to a rigid click track or drum machine, not a human drummer. Fortunately, when controlling effects, synchronization *to a particular tempo* is not an issue. This is because a sequencer can serve simply as a way to trigger program changes and spew out MIDI controller data on cue. You do need to sync the sequencer to tape, but it need not correlate to a tune's tempo.

To understand why this is so, we need to distinguish between *absolute* time and *musical* time. Musical time is relative; a quarter note has a different duration at 60 BPM compared to that same note at 140 BPM. Absolute time is a constant. In other words, if you're 1 minute and 12 seconds into a song, it doesn't matter what the tempo of the tune may be, or even if there is a tempo. You're simply marking a point that's 1 minute and 12 seconds into the tune, which will always correspond to a particular place in the music if the sequencer plays back at a consistent tempo.

Most of the time, sequencers are used in the context of musical time. For example, you might consider a program change or other MIDI command as occurring at the beginning of the 43rd measure of a tune. However, *by keeping a constant tempo* we can also treat the sequencer as an absolute time device that simply identifies elapsed time since the beginning of the tune. So, as long as the sequencer is synched to tape, if we record a piece of MIDI data into the sequencer at a particular time, it will always play back at that particular time.

One reason it's not necessary to follow a particular tempo is because sequencers identify time with a great deal of precision. A typical computer-based sequencer running at 180 BPM can divide time into approximately 1,000 discrete events *per second.* Think of this as 1,000 "memory slots" into which the sequencer can record data, which means the sequencer can record a piece of MIDI data about every 1/1000th of a second (1 ms).

Suppose you want to record a program change command at the beginning of measure 43. The odds are excellent that there will be a corresponding "memory slot" that falls almost, or maybe even exactly, on that beat. Worst case, though, is that the command will be issued within ±0.5 ms of wherever the musical beat falls—definitely close enough.

At the Session

We'll assume that the rhythm section has already cut its tracks, and it's time for you to do some overdubs or automate some signal processing changes. First, gather your tools:

- MIDI sequencer
- Sync-to-tape box or interface (if needed to synchronize the sequencer to tape)
- Some way to record program changes and continuous controllers into the sequencer (such as a MIDI fader box, MIDI footswitch/pedal combination, or a keyboard that can generate the desired types of MIDI data; see sidebar)

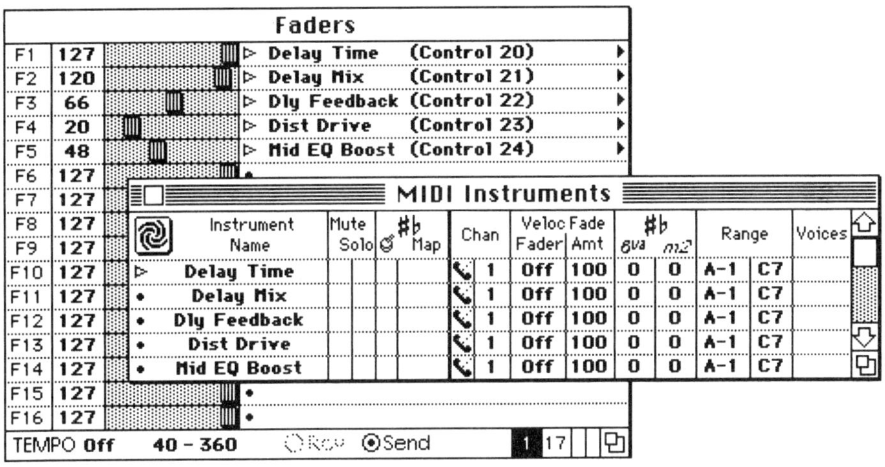

Fig. 6-1: The faders in Opcode's Vision program. The front window assigns the fader to various channels, ports, and ranges; the rear window shows the faders themselves and their controller assignments. With Vision, clicking on the rightmost triangle in a fader strip calls up a popup menu that contains a list of all available controllers.

Now that you've collected the gear, here's the general procedure on how to use it. We'll assume you want to automate program changes and continuous controller messages.

MIDI Data Generators

Many devices can generate MIDI data; patching the MIDI out from one of these devices to the sequencer's MIDI in lets you record MIDI data. Here are the most common data generators:

• **MIDI keyboard.** A typical MIDI keyboard can generate program changes and a variety of other control signals such as pitch bend, pressure, mod wheel (controller 01), footpedal (controller 04), etc. Not all keyboards can generate all MIDI commands (nor can all on-board keyboard sequencers record all MIDI commands). However, some advanced keyboards and master controllers have dedicated "data sliders" that can be set to any controller number.

• **Footswitch / footpedal combination.** Guitarists who use MIDI footswitches and pedals can generate control data directly from these. The footpedal may or may not be assignable to different controller numbers; if not, the sequencer can often convert data recorded as one controller number into a different controller number.

• **MIDI fader box.** This is a compact box that consists of a number of slide faders, each of which can be assigned to generate a particular type of MIDI data, such as different controllers over one or more MIDI channels. This type of device is well-suited to generating MIDI data since you can move several faders at once, or grab a particular fader to generate a particular controller.

• **Software MIDI faders.** Several sequencers include on-screen "virtual faders" that you can assign to various controllers (Fig. 6-1). In this example faders F1-F5 are assigned to controllers 20-24, which all transmit over channel 1. To help remember which fader controls which function, each has been named in the MIDI Instruments window.

These "fader" motions can be recorded as part of the sequence, or sent in real-time to the MIDI output. On playback, with most programs the faders move to give a visual indication of the data value (this also looks very cool).

1. Patch the MIDI data generator's MIDI out to the multieffects' MIDI in so that you can hear the results of the continuous controller and program changes (Fig. 6-2).

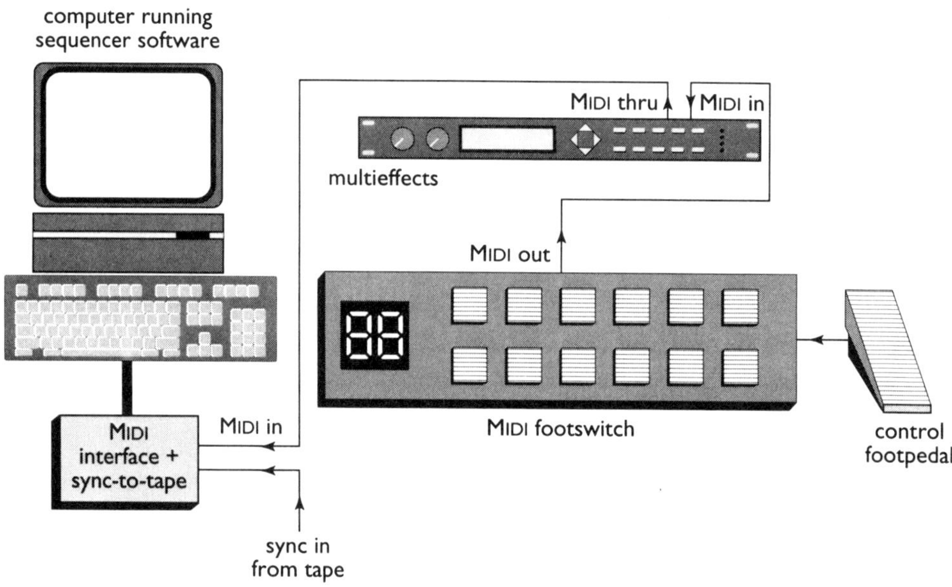

Fig. 6-2: Hookup for recording MIDI data into a sequencer. The MIDI data generator is a footswitch and footpedal, as typically used by guitarists. Alternately, you could use one of the other types of generators mentioned in the sidebar.

2. Patch the MIDI generator's MIDI thru (which transmits a copy of the data appearing at the multieffects' MIDI in) to the sequencer's MIDI in.

3. On an empty tape track, record a sync track to which the sequencer can synchronize.

4. Set the sequencer to some arbitrarily fast tempo, such as 180 beats per minute. Since the tempo probably won't be following the track, turn off any click or metronome option.

5. Start the tape. Record your controller and program changes into the sequencer (which should be synched to tape) while listening to the track. If you're using a MIDI generator such as a continuous controller pedal and are playing along at the same time, don't worry too much about your playing—just get the controllers recorded in the right place at the right time.

6. Prior to playback, disconnect the MIDI generator's MIDI out from the multieffects' MIDI in; patch in the sequencer's MIDI out instead (Fig. 6-3).

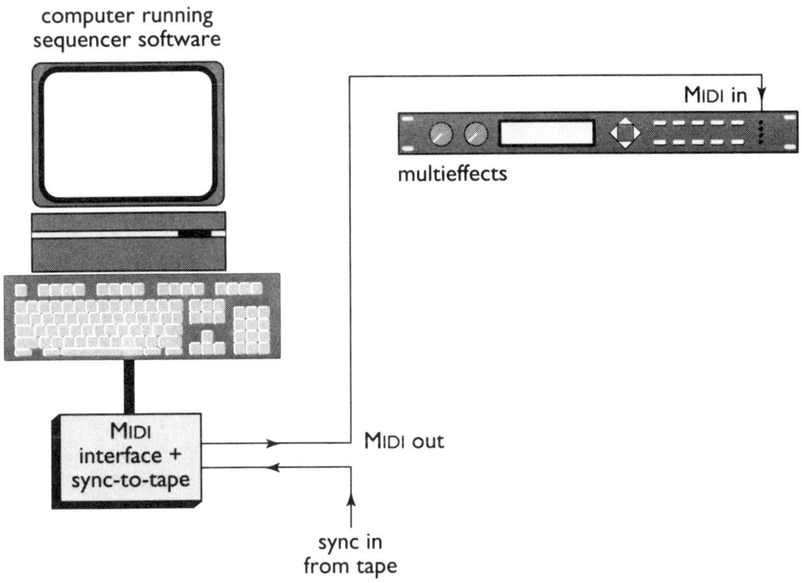

Fig. 6-3: Hooking up a multieffects and sequencer for sequence playback.

7. Put the tape into play mode. After a second or two, the sequencer will autolocate to the correct place on the tape and play back any MIDI messages exactly as you recorded them. If you need to overdub or fix a part, you can do so without having to worry about footswitches, faders, pedals, or other distractions.

That's all there is to it. Also remember that one of the advantages of using a sequencer is that you can edit the data. For example, if a signal processor is slow to respond to program changes, use the sequencer's "track shift" (or equivalent) command to shift the program changes slightly ahead of the beat.

Potential Problems With Sequencing

Programming lots of controller messages into a sequence uses up a considerable amount of memory. Moving a mod wheel, pedal, or fader sprays out hundreds of events; too many of these can cause timing delays or result in "MIDI clog," where the sequencer simply gives up in frustration. Solving this problem requires deleting unneeded data.

Fortunately, many times you can use a "snapshot" approach and just insert a single controller value at strategic points. The parameter will remain at this value until changed. Figs. 6-4A through 6-4C show a continuous controller doing a fade in for the master volume parameter. Fig. 6-4A shows the fade entered with a pedal; Fig. 6-4B shows an edited "snapshot" version of the fade. There is little audible difference between the two, but Fig. 6-4B uses much less data.

Some sequencers include data thinning algorithms that reduce the amount of MIDI data automatically, which saves editing time. Fig. 6-4C shows the same signal, but thinned using a sequencer's data thinning algorithm that removes data falling within a specified number of clock pulses, or within certain values compared to a neighboring piece of data. Clearly, B and C save a lot more memory compared to A.

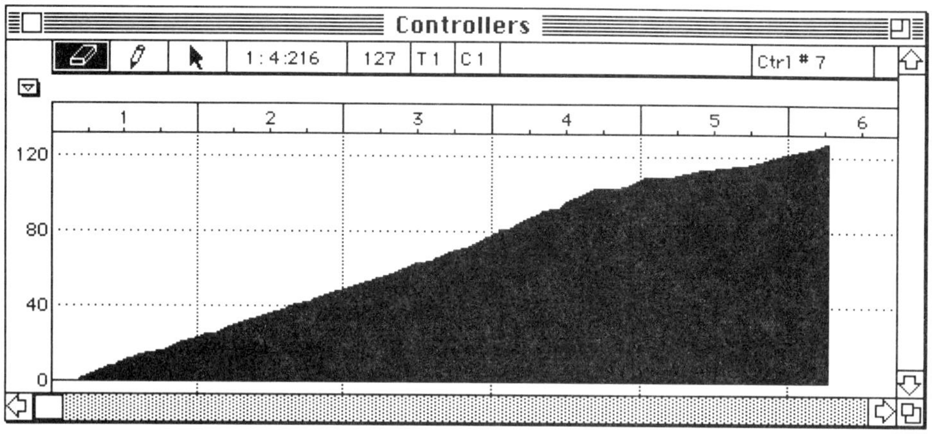

Fig. 6-4A: Controller events for a fadein, recorded from a MIDI *pedal assigned to volume.*

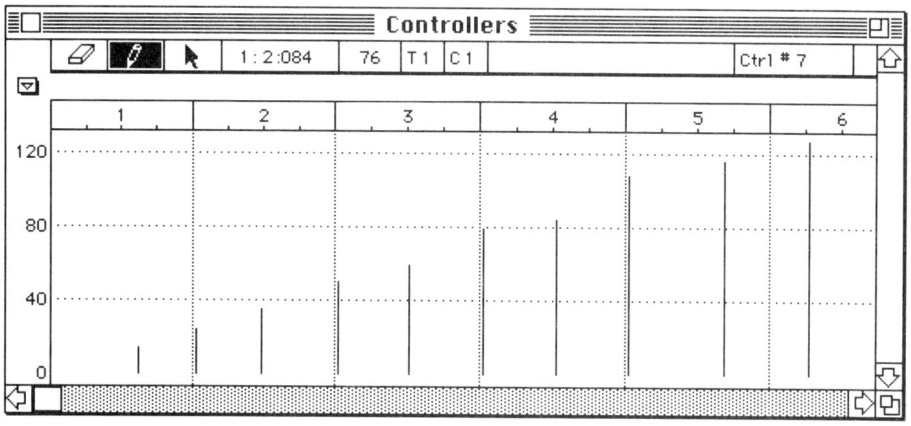

Fig. 6-4B: The controller events from Fig. 6-4A converted to "snapshots."

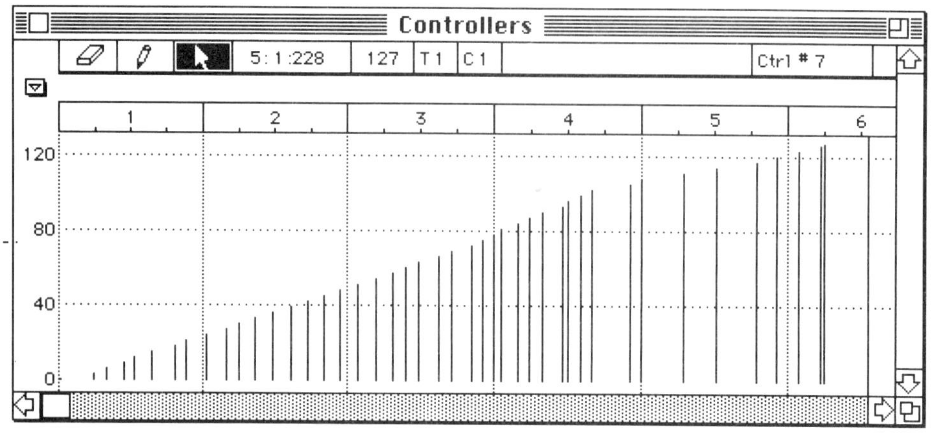

Fig. 6-4C: The controller events from Fig. 6-4A after thinning by a sequencer's thinning algorithm.

Although in most cases you won't have to resort to editing your data, it's important to remember that sequencers do have limitations. If you're trying to use lots of controller data and experience timing problems, dig into the data stream and do some editing.

Automated Vocal Mixdown

A typical multieffects has EQ, time delay, compression, and the ability to control overall volume with MIDI controller 7. If that sounds like an ideal combination for automated mixing of vocal tracks without having to spend megabucks on a completely automated console...well, it is.

Run a vocal tape track directly into the multieffects. If you don't need to use a mixer's bells and whistles (the multieffects already has plenty of those), the multieffects can patch into a mixer effects return, thus freeing up an additional mixer channel as part of the deal.

Use the multieffects processing to give the sound you want, and continuous controllers driven by a sequencer to control the overall volume (and/or make changes to the effects).

Automating Effects in Live Performance

Automating effects changes can also work "live" if you play to a MIDI sequencer or drum machine and use defined song structures. Automation gives an incredible amount of freedom, since all you have to do is play and sing instead of do the "footswitch/pedal tap dance."

(Incidentally, devices such as the Aphex Studio Clock can "listen" to the drummer and generate a sync track suitable for driving a sequencer. This lets you do the tricks we're about to describe during live performance without having to follow a drum machine or a sequencer's internal clock. However, this is a somewhat riskier process, as it requires a relatively consistent drummer.)

As one example, a few years back I played in a synth duo. We recorded the drum sounds into a sampling keyboard, then sequenced them into drum parts using the keyboard's onboard sequencer. I also recorded all continuous controller and footswitch changes for my guitar and vocal processors into the sequencer.

Perhaps the most interesting thing about doing this is that with a "three piece" band like mine (two humans and robodrums), there's a tendency for the sound to be a bit thin. One option is to sequence synthesized background parts like strings and keyboards, but then the audience can tell something phony is going on. By automating a series of signal processing changes, the guitar and vocal sounds have much more variety and interest. There's no need to add more instruments if you can make the most out of the ones you already have.

Automated Multieffects Limitations

It's amazing what technology can do these days, but nothing is perfect—and that includes multieffects. Here are some problems you may run into, with suggested solutions.

Program Change Problems

When a multieffects switches from one sound to another, it has to "flush" its memory and load in new parameters, which can take several milliseconds. Different units handle this in different ways—some mute the signal, some bypass the processed sound temporarily, and some just make nasty glitching noises.

What's more, the amount of glitching can differ from program to program. For example, programs that use a lot of time-based processing (delay, pitch transposition, etc.) are more likely to glitch than ones that vary amplitude or frequency response (EQ, compression, etc.). Glitching is an inevitable aspect of programmability and while some units are better than others, for the foreseeable future effects will stutter a bit when transitioning between some sounds.

One solution: if you mostly use a clean or distorted sound with effects added in the background, run the main sound through one amp or mixer channel and the multieffects through a second channel. If there's a glitch while the effect changes, you'll continue to hear your playing through the main channel, which will help mask any problems.

If you're automating program changes with a sequencer and experience really nasty glitching, try connecting a MIDI-controlled mixer or attenuator (such as the Niche ACM or JL Cooper's MixMate) after the effect's audio output, and program a mute while the program changes. Again, this works best if you have some straight signal playing along.

Continuous Controller Problems

Some continuous controllers glitch when altered, and some don't. Parameters that control time tend to glitch the most (pitch transposition amount, delay time, and the like). Parameters that control amplitude glitch the least. Parameters that control filters are sometimes clean, and sometimes dirty—it depends mostly on the number of steps into which a parameter is quantized (as explained in Chapter 2). An analog control has an essentially infinite number of steps; to replicate that digitally costs a fortune, so most parameters will be quantized into anywhere from 2 to 256 (or sometimes more) steps.

The more steps, the more continuous the control, and the smoother the feel. If you sweep through a control that's quantized into few steps, you'll hear each step go by—a phenomenon popularly referred to as "zipper noise," since the resulting sound gives a slight glitching as you go through each step.

There are two main solutions for dealing with continuous controller glitching:

- Avoid playing when changing the value of a glitch-prone parameter

- Use a snapshot approach (instead of doing continuous sweeps) when a parameter needs to jump from one value to another. This doesn't solve all types of controller glitching, but it's worth a try.

Eliminating Program Change Glitches Entirely

When cost is no object, there is a way to avoid program change glitches: get two multieffects and fade between them with a pan pedal or MIDI commands. While effects device A is active, switch programs on effects device B (which you can't hear) before you actually need to change programs. When it's time to switch, use the volume pedal to fade out device A and fade in device B. Now you can change effects on device A while device B is selected; to change programs again, fade out device B and fade in device A.

There are several ways to accomplish this, but the simplest is to assign each multieffects to a different MIDI channel so that each device can receive its own program changes (you can also change the effects manually), then use an audio panning footpedal to crossfade the audio outputs. If you have a standard, passive stereo volume pedal that uses a dual-ganged potentiometer, you can modify it into a "crossfader" (see Fig. 6-5).

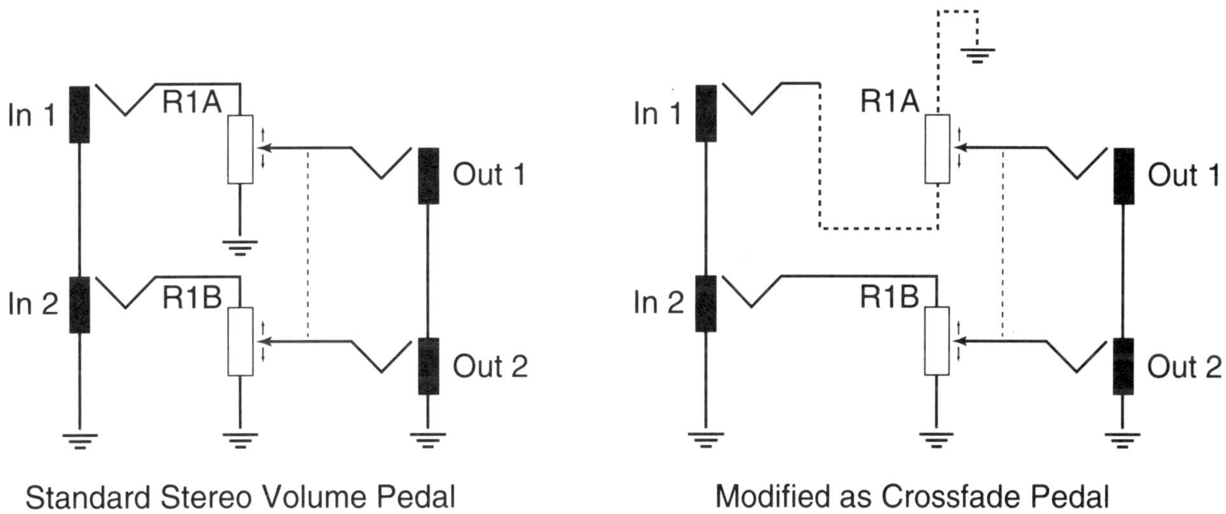

Fig. 6-5: How to rewire a standard stereo volume pedal into a crossfade pedal. The basic idea is to reverse the "hot" and "ground" leads of one of the ganged potentiometers (but not the other). As R1a fades out, R1b fades in.

An alternative to buying two effects is to use continuous controllers to change parameters within a *single* program. While more limited, suppose you have a setup where distortion and delay are in series, and you want to switch between a clean sound, a distorted sound, a clean sound with tight echoes, and a distorted sound with long echoes. Assign the distortion amount, delay time, and delay mix to three separate continuous controllers, and vary their values with control pedals or from a sequencer. Chances are changing these parameters will glitch less than what you'd get from changing programs (especially if you don't play while the delay time parameter is changing).

Processing Acoustic Guitar

Multieffects are not just for electric instruments. With more and more acoustic guitars including pickups, multieffects can do a lot to enhance the sound of acoustics, whether live or in the studio. The object here is not to make an acoustic guitar sound like an electric, but to get exceptional acoustic sounds.

We'll assume that your acoustic has been electrified and can generate a signal of sufficient level, and of the proper impedance, to drive a multieffects. If you're not sure about this, contact the manufacturer of the pickup assembly, or whoever did the installation.

Improving Tone and Reducing Feedback

Most electrified acoustics have frequency response anomalies—peaky midrange, boomy bass, and so on—caused primarily by the interaction of the guitar body, the strings, and the pickup. While some of these anomalies are desirable (classical guitars wouldn't sound as full without the bass resonance most instruments exhibit), some are unwanted. Smoothing out the response is a task for equalization.

As with electric guitars, graphic EQ is best for general tone-shaping—making the sound "brighter" (more treble), "warmer" (more lower midrange), "fuller" (more bass), etc. A parametric equalizer excels at solving specific problems, such as boominess or thinness at a particular frequency.

Either type of equalization can help balance your guitar with the rest of the instruments in a band. For example, both the guitar and the male voice tend to fall into the midrange area, so they compete to a certain extent. Reducing the guitar's midrange response will leave more "space" for vocals.

Parametrics can be helpful in taming feedback. Feedback relates to level, and since a guitar body's resonances increase the level at certain frequencies, feedback will occur at the frequencies exhibiting the strongest resonances. As you turn up the volume to the point of feedback, you'll hear a whining tone. Set the parametric for a narrow bandwidth and cut the response by 6 dB or so, then vary the frequency control until the whine goes away, indicating that you've found the particular resonant frequency causing the feedback. You should now be able to increase the volume a bit until feedback kicks in from the next most prominent resonant frequency, at which point you can similarly adjust another parametric stage to reduce feedback even further.

A parametric can also be indispensable for getting more level on tape from classical guitar. Classical guitars tend to have strong resonant peaks, usually in the bass and lower midrange, that make certain notes, or even entire frequency ranges, jump out. This is a particular problem in the studio: if you set the record level low enough to accommodate the peak, the other notes will sound weak by comparison. If you increase the record level, though, you run the risk of creating distortion whenever the guitar plays a note in the range of the peak.

Limiting or compression will tame the peak, but only at the expense of a somewhat squeezed sound. A more natural-sounding option is to use a parametric to counteract the effect of the peak.

To find the peak frequency, turn down the monitoring system volume, and set the parametric's boost parameter to full, with a moderate bandwidth. Play the guitar and vary the frequency control; when you hit the right frequency range, the volume will jump dramatically. Now change the boost to a cut. This will tame the peak and let you turn up the record level higher, thus putting a higher average signal level on tape.

Brightness or Fullness Without Equalization

Using pitch transposition to transpose an acoustic guitar sound up an octave (for a brighter sound) or down an octave (for a fuller sound) can sound pretty good, providing that you mix the transposed signal well in the background of the straight sound—you don't want to overwhelm the straight sound, particularly since the processed sound will generally sound artificial anyway.

Bigger Sounds

Adding some delay can create a bigger-than-life, ensemble sound if you set a short delay (50 to 100 milliseconds) and turn the feedback (or regeneration) and modulation controls to minimum (or optionally, add a little modulation for a chorusing effect that "thickens" the sound). This produces a slapback echo, and gives a tight doubling effect.

Increased Sustain

The guitar is a percussive instrument that puts out a huge burst of energy when you first pluck a string, but then rapidly decays to a much lower level. If you want more sustain, limiting is the answer. Don't set the threshold too low, or the guitar will sound "squeezed" and unnatural, and you may experience feedback problems. Also, don't use compression instead of limiting. A compressor tries to maintain a constant output in the face of varying input signals by not only attenuating high-level signals, but by amplifying low-level signals. When playing live, this can increase the potential for feedback.

Pedaling Your Way to Bigger Sounds

If you have a two-channel amp or mixer, one trick that's applicable to all of the above options is to split your guitar signal into two paths with a Y-cord. One split carries the straight guitar sound, while the other goes through a volume pedal before feeding the multieffects. Use the volume pedal to mix in as much processed sound as you want.

The possibilities for processing acoustic guitar are just as exciting as for processing electric guitars. The best way to learn, though, is by experimentation. You never know what sounds you'll discover as you plug your guitar output into a multieffects and start editing parameter values.

In Seach of the Big Electric Guitar Sound

Many guitarists, whether they play live or in the studio, are looking for ways to produce the Ultimate Electric Guitar Sound—a sound so full, so ringing, so resonant, and so beautiful that fans and record company execs alike will quiver in awe upon hearing it. Well, we're not quite there yet. Meanwhile, there's still plenty you can do to make your guitar sound larger-than-life; here are some favorite tips, including a few that don't involve using a multieffects.

Playing Technique

Much of getting a big sound depends on playing technique. Chord voicings that use open strings, or are played low on the neck, help give a more ringing sound. Thicker gauge strings (top E string = 0.010″) also give a meatier sound—as well as stand up better to repeated string-bashing.

Playing intervals *(i.e.,* octaves à la Wes Montgomery, or parallel fourths/ fifths) can help make your single note solos sound bigger as well (if you're not up to doing this with your fingers, you can always cheat and use the pitch transposition option in your multieffects). Certain guitar adjustments, such as making sure the pickups are properly positioned with respect to the strings, can also contribute toward a bigger sound.

Perhaps the most popular route to a big guitar sound is to use a multieffects, which can alter three major sound parameters that help create a "big sound": tone quality, sustain, and ambience (room acoustics).

Tone Quality

Electric guitars usually generate more low-frequency audio energy than high-frequency audio energy. Therefore, boosting high or upper midrange frequencies (starting at approximately 3 kHz) can "fill out" the sound of an electric, giving it some extra brightness that makes it sound bigger and also increases articulation.

If you have a two-stage parametric, try boosting the very low end (for a feeling of "power") as well as the higher frequencies. Better yet, simply cut some of the midrange response and increase the volume a bit at your amp. This will accomplish the same result *(i.e.,* accentuate the high and low frequencies), but since you're cutting rather than boosting audio energy, you run a far lower risk of encountering distortion.

Another way to alter tone in the studio is by overdubbing multiple guitar parts. However, the more overdubs, the greater the risk of producing both frequency response cancellations and a "blurring" of the sound (although of course, sometimes this is desirable). It's often better to use a multieffect's delay or pitch transposer modules to "synthesize" the extra part and create a stereo image.

Sustain

While compressors add sustain, if you overcompress your guitar it will actually sound thinner. Always use compression very judiciously if you want the biggest possible sound.

Ambience

It seems axiomatic that all guitar amps include a reverb, and why not—reverb was one of the first effects recognized as giving a "big" sound. Other time delay devices (flangers, chorus units, echo units) can also create ambience; however, these devices must be used with extreme care, since they can just as easily diffuse a sound as augment it. For example, a flanger or chorus puts lots of notches (dips) in a guitar's frequency response, which may dilute the sound somewhat.

Here We Go Loop de Loop, Interfacing Guitar Amps and Multieffects

We've all seen those loop jacks on an amp's back panel, but what do they *really* do? And why does plugging an effect into those jacks sometimes produce either unbearable noise or distortion? Some multieffects are better suited to interfacing with guitar amps than others; here's why.

A Little History

The problems associated with interfacing guitars to electronic devices all started because standard pickups put out very low-level signals. Because the signal-to-noise (S/N) ratio of any electronic device contrasts the level of signal with the amount of residual noise (which is fairly constant), higher level signals lead to better S/N ratios (providing the signal level isn't so high that it leads to distortion).

There are two main types of multieffects used by guitarists: units specifically designed to work with the low-level signals produced by guitars, and studio (often rack-mount) effects designed to work with higher-level (called *line* level) signals favored by tape recorders and mixers. The former type of device can simply insert between your guitar and an amp, since everything's matched: the low-level guitar signal feeds an input that's designed to accept low-level signals.

Insert a line level device between the guitar and amp, and you have problems. First, the effect will expect to see a high level signal, which the guitar doesn't provide. Second, the device will probably put out a fairly hot signal, which will more than likely overload the amp input. Fortunately, there is a solution.

Enter the Effects Loop

Within a guitar amp, the guitar signal first goes through a preamp to bring the level up sufficiently to drive the power amp. This preamplified signal is strong enough to drive line-level effects units, so somewhere along the line someone got the idea to insert a couple of jacks between the preamp output and power amp input (usually called the send and return jacks respectively). Switching jacks are usually used so that with nothing plugged in, the signal goes directly from the preamp out to power amp in. Plugging into the return jack breaks this connection.

The levels at this point in the circuit are ideal for driving line-level effects; devices that accept only guitar-level signals can still patch between the guitar and amp. As a bonus, the send provides a post-preamp signal that can patch to a slave amp or recording mixer input.

Unfortunately, there are some complications. Some loops are designed not with multieffects in mind, but assume that you'll use guitar-level oriented effects pedals. This type of loop may not provide enough juice for devices that want to see line levels; the multieffects' high output might also overload the power amp. A multieffects with individually adjustable input and output controls is your best bet for interfacing with an effects loop, since you can optimize the input and output levels to accommodate high- and low-level loops. However, there are other problems to consider—what if the multieffects is stereo, and the loop is mono?

Guitar amp designers are a creative lot, and they've come up with a variety of designs to tackle these problems. Since they don't know exactly what type of effects people will be using, they've come up with a variety of compromise solutions.

- **Dual loops.** Amps with this design include a low-level loop (typically pre-EQ) and high-level loop (typically post-EQ, between the preamp and power amp). You'd insert the multieffects into the high-level loop.

- **Switchable loop gain.** This approach adds a two-position switch to the standard set of loop jacks. One switch position provides the levels necessary to interface with low-level effects, while the other position conditions the levels for line-level effects. Use the line-level position and adjust the multieffects' input and output level controls accordingly.

- **Adjustable loop level controls.** This is a more flexible version of the above scheme, but now you have to mess with both the loop *and* multieffects input and output controls to find the right setting.

- **Dual channel loops.** Some two-channel amps provide a single effects loop that affects both channels; others provide individual loops for each channel, which seems preferable since you often want each channel to have its own distinctive sound.

In today's high-tech arena, loops are constantly being improved. Some offer bypass switches, some accommodate stereo effects or at least accept a stereo return, some send the effect signal in parallel to the normal signal path (rather than interrupt the straight signal) so you can set a precise mix of dry and processed signal, and so on. In programmable amps and preamps, the loop status (bypass or active) is often programmable as well.

With so many variations it's somewhat risky to generalize, but the bottom line is that a multieffects unit can often be adjusted to work as either a low-level or line-level device. Nonetheless, you'll probably get the best results if you treat the effect as a line-level device and use the line-level loop jacks. If this creates distortion and no amount of juggling the multieffect's input and output controls provides a solution, then insert the multieffects unit into the low-level loop jacks (if present) or as a last resort, between the guitar out and amp input.

All of this may sound complicated, but in practice, most of the time when you plug an effect into a set of loop jacks everything will work as expected. Should you encounter difficulties, though, a little trial-and-error, coupled with the background theory presented above, should go a long way toward solving the problem.

Retrofitting Standard Pedals to Generate Continuous Controllers

A MIDI continuous controller pedal looks like a standard volume pedal, except that it has a MIDI output jack. Several companies make such pedals, but as of this writing there's only one device that can retrofit *existing* pedals and footswitches for this application: Anatek's Pocket Pedal (PP for short).

The PP is a small box with two input jacks (one for a standard footpedal, one for a standard footswitch) that converts pedal motion and footswitch presses into continuous controller commands that appear at the PP's MIDI out jack. You then patch this into the effect's MIDI in. You also need to connect the effect's MIDI out to the PP's MIDI in, since the PP draws its power from the MIDI out line.

The footpedal/PP combo can send over any combination of controllers 1, 5, 7, or pitch bend. The footswitch sends over controllers 64, 65, and 66, which are usually used for on/off control applications. Data is normally sent over MIDI channel 1, but if programmed via a MIDI keyboard, guitar controller, or other MIDI note generator, the PP can send over any combination of (or all) MIDI channels.

The PP works with virtually any resistive footpedal (stereo or mono jack) including the Yamaha FC-7, Korg EXP-2, any Ensoniq or E-mu keyboard pedal, Roland EV-5, DOD/DigiTech volume pedal, Ibanez VL10, etc. About the only requirement is that the maximum potentiometer resistance shouldn't exceed 300 kohms. The footswitch can be either normally open or normally closed; as soon as you plug in the pedal and/or footswitch, the PP analyzes them and adapts its response to that of the devices being used.

If you already have a standard pedal, the PP is a simple way to start experimenting with continuous controllers. You might be surprised at what this little box will let you do.

The Top 10 Pedal Targets

Controlling effects parameters via footpedal can add real-time expressiveness to your playing. The trick is then to figure out which parameters you want to control to obtain particular effects. Some multieffects make it easy: they offer companion footpedals and have patches pre-programmed to work with those pedals. But this isn't always the case, and besides, the manufacturer's idea of what you want to do may not be the same as what you want to do.

Certain parameters are a natural for foot control; here are ten that can make a big difference to your sound.

- **Distortion drive.** This one's for guitar players. Most of the time, to go from a rhythm to lead setting you step on a switch, and there's an instant change. Controlling distortion drive with a pedal lets you go from a dirty rhythm sound to an intense lead sound over a period of time. For example, suppose you're playing eighth-note chords for two measures before going into a lead. Increasing distortion drive over those two measures builds up the intensity, and slamming the pedal full down gives a crunchy, overdriven lead.

- **Chorus speed.** If you don't like the periodic whoosh-whoosh-whoosh of chorus effects, assign the pedal so that it controls chorus speed. Moving the pedal slowly and over not too wide a range creates subtle speed variations that impart a more randomized chorus effect.

- **Echo feedback.** Long, languid echoes are great for accenting individual notes, but get in the way during staccato passages. Controlling the amount of echo feedback lets you push the number of echoes to the max when you want really spacey sounds, then pull back on the echoes when you want a tighter, more specific sound. Setting echo feedback to minimum gives a single slapback echo instead of a wash of echoes.

- **Echo mix.** Here's a related technique where the echo effect uses a constant amount of feedback, but the pedal sets the balance of straight and echoed sounds. The main differences compared to the previous effect are that when you pull back all the way on the pedal, you get the straight signal only, with no slapback echo; and you can't vary the number of echoes, only the relative volume of the echoes.

- **Graphic EQ boost.** Pick one of the midrange bands between 1 and 4 kHz to control. Adjust the scaling so that pushing the pedal all the way down boosts that range, and pulling the pedal all the way back cuts the range. For solos, boost for more presence, and during vocals, cut to give the vocals more "space" in the frequency spectrum.

- **Reverb decay time.** To give a "splash" of reverb to an individual note, just before you play the note push the pedal down to increase the reverb decay time. Play the note, and it will have a long reverb tail. Then pull back on the pedal, and subsequent notes will have the original, shorter reverb setting. This works particularly well when you want to accent a drum hit.

- **Pitch transposer pitch.** For guitarists, this is almost like having a "whammy bar" on a pedal. The effectiveness depends on the quality of the pitch transposition effect, but the basic idea is to set the effect for pitch transposed sound only. Program the pedal so that when it's full back, you hear the standard instrument pitch, and when it's full down, the pitch is an octave lower. This isn't an effect you'd use everyday, but it can certainly raise a few eyebrows in the audience as the instrument's pitch slips and slides all over the place. By the way, if the non-transposed sound quality is unacceptable, mix in some of the straight sound (even though this dilutes the effect somewhat).

- **Pitch transposer mix.** This is a less radical version of the above. Program the transposer for the desired amount of transposition—octaves, fifths, and fourths work well—and set the pedal so that full down brings in the transposed line, and full back mixes it out. Now you can bring in a harmony line as desired to beef up the sound. Octave lower transpositions work well for guitar/bass unison effects, whereas intervals like fourths and fifths work best for spicing up single-note solos.

- **Parametric EQ frequency.** The object here is to create a wa pedal effect, although with a multieffects, you have the option of sweeping a much wider range if desired. Set up the parametric for a considerable amount of boost (start with 10 dB), narrow bandwidth, and initially sweep the filter frequency over a range of about 600 Hz to 1.8 kHz. Extend this range for a wider wa effect if desired. Increasing the amount of boost increases the prominence of the wa effect, while narrowing the bandwidth creates a more intense, "whistling" wa sweep.

- **Increasing the output of anything** ***(e.g.,*** **input gain, preamp, etc.) before the compressor.** This allows you to control your instrument's dynamic range; pulling back on the pedal gives a less compressed (wide dynamic range) signal, while pushing down compresses the signal. This restricts the dynamic range and gives a higher average signal level, which makes the sound "jump out." (You've heard this effect before with TV, where commercials are compressed to make them sound louder than the program you're watching.)

  This effect is possible only if you can control the output level of the stage prior to the compression.

  Control the level so that pushing the pedal down increases the amount of signal feeding the compressor. When you push down on the pedal, the dynamics will change so that softer playing will come up in volume. This can make a guitar seem more sensitive, as well as increase sustain and make the distortion sound smoother.

Retrofitting Older Amps and Effects for MIDI Control

Several amplifiers and older effects have phone jacks that accept a footswitch for functions like channel switching, reverb in/out, bypass, tremolo on/off, and the like. These can be retrofitted for MIDI control with an accessory device that turns program changes or continuous controllers into switch closures.

Fig. 6-6 shows the interface for this type of device. There will be two terminals that act as a MIDI-controlled switch, usually labeled + and - (or hot and ground). You wire these two terminals to a phone plug that inserts into the amp or effects footswitch jack.

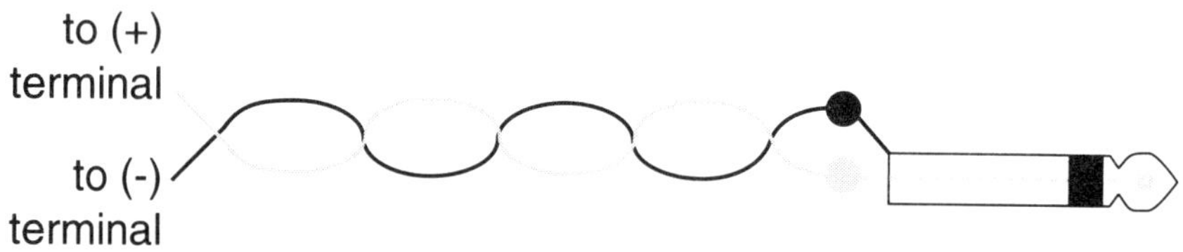

Fig. 6-6: The hardware needed to interface a MIDI-controlled switch closing accessory to a standard footswitch jack.

There are two types of switches: transistor or relay. With transistor switches, the hot/ground polarity is important—these terminals must match up with the plug hot and ground connections. A relay is a more universal option because it doesn't care which terminal goes to ground and which is hot.

As of this writing, suitable products are made by Scholz R&D and Lake Butler Sound. An alternative is the MIDItools Computer running project #8 (this device is described fully in *Digital Projects for Musicians,* published by Amsco). Project #8 provides four relays that can be triggered by MIDI continuous controller commands. Each is assigned its own continuous controller; values of 64 or over open the relay, and values under 64 close the relay.

Footswitch Patch Organization for Live Performance

It takes some forethought to arrange your MIDI footswitch programs for the most efficient program access when playing live. This is because many MIDI footswitches use a "bank/program" protocol that arranges programs as several banks; switches within each bank select individual programs. Banks typically have anywhere from 5 to 10 switches.

- If you're into changing programs a lot, consider dedicating each bank to a particular song, and select programs for that song within the bank. Each required sound will be only one footswitch press away, and you'll have plenty of time between songs to change banks. Typical high-end footswitches usually offer at least 16 banks of eight programs, which should be enough for the average set. Between sets, you can load new data with a system exclusive storage device (as described at the end of Chapter 5).

- If you use only a limited number of programs, it's probably best to assign the programs you need to switches within a single bank. As long as you remain in that bank, accessing any sound requires only a single footswitch press. If you use more sounds and spread them over multiple banks, you'll need additional footswitch presses—at least one to select the bank, and another to select the program.

- Some footswitches include *increment* and *decrement* switches to select the next higher-numbered program or lower-numbered program, respectively. If you have a fixed set list, you can program the various sounds sequentially—*e.g.*, the first sound you need is program 1, the second is program 2, and so on. Then, whenever you need to move on to the next patch, press the increment switch. The main disadvantage is that this wastes memory slots in your multieffects if you use the same program a lot, since you'll have to copy the same sound to multiple memory locations.

- The easiest option is to have a sequencer do all the dirty work for you and select not only programs, but vary parameters in real time as well. This type of "automated effects system" works well both live and in the studio (see "Totals Effects Control in the Studio" and "Automating Effects In Live Performance" toward the beginning of this chapter).

Using MIDI Guitar to Control Signal Processors

Pedals aren't the only option for varying parameters; here's a particularly intriguing alternative. A MIDI guitar that uses pitch-to-MIDI conversion (like the various Roland models) can be played as a standard electric guitar. Normally the MIDI data drives synth modules, but there's no reason why you can't use the MIDI data solely to control processors that affect the straight guitar sound. Here are just a few suggested applications:

- Tie echo feedback to note pitch, so that higher notes give longer echoes (good for solos).

- Have EQ track the notes being played so that lower notes are bassier and higher notes have more treble.

- Control chorus depth with velocity *(i.e.,* the dynamics of your playing) so that louder sounds "swirl" more.

- Use the guitar's "whammy bar" or other physical controllers (knobs, switches, etc.) to vary signal processor parameters in real time from your axe.

Programming Multieffects Via Computer

Some multieffects are becoming almost as complex to program as synthesizers. Not only do modern signal processors include a slew of programmable parameters, but also, many parameters can now be controlled in real time via MIDI continuous controllers or other control signals.

Unfortunately, your access to these parameters is often a small LCD or LED display, with editing done through the tedious method of selecting a parameter, altering its value, selecting another parameter, altering its value, and so on. Although there are quite a few computer-based synthesizer editing programs for various computers that let you program synth sounds on-screen, there aren't a lot of editors for signal processors.

This point was driven home the other day as I slogged through programming the graphic EQ parameters on a digital multieffects unit. Not only did it take a lot of button presses to switch among parameters and alter values, it was frustrating to vary just one parameter at a time, listen to the results, vary another parameter, listen again, go back and re-tweak previous settings...in fact, you're probably dozing off just listening to me describe the editing procedure. Well, it's even more boring when you're trying to tweak up 30 or 40 patches.

However, some MIDI sequencers can provide a relatively painless way to program multieffects parameters since they offer on-screen "virtual faders." Each fader (there are usually several faders available at one time) can be programmed to output a specific continuous controller. Before putting these to use, however, we need to cover a bit of theory as to how signal processors handle continuous controllers.

A Typical Example

Probably the best way to illustrate how this technique works is to choose a specific example, so we'll program a digital graphic equalizer. Note that there can be significant variations between different multieffects in how they display your edits; for example, sometimes a device's display will update a parameter's value as it's changed, and sometimes it will always show the programmed value even though the parameter is being varied via MIDI continuous control.

Graphic EQ Control

The first step is to define which controllers affect which parameters. This involves dialing up each graphic EQ band parameter and linking it to a specific controller. In this example we'll assign the 63 Hz band to controller 10, the 125 Hz band to controller 11, the 250 Hz band to controller 12, etc. Remember that some effects use a global control approach and some assign controllers on a per-program basis (see Chapter 5 for information on the difference between "global" and "per-patch" controller assignments).

The next step is to program the sequencer's on-screen faders to match the multieffects controllers you want to vary. Virtually all "major league" sequencers include some kind of programmable virtual fader option, and let you save the fader assignments either by themselves or as part of a sequence (of course, you needn't program anything into the sequence if you just want to use the faders).

Fig. 6-7 shows the Instruments window for an early version of Deck's "Metro" sequencer, set up for graphic EQ control. Note that each fader has been named with the frequency band it controls; controller assignments themselves are set in a window that pops up when you click on the fader name. They're all set to transmit on channel 1 (although usually, it makes sense to set the signal processor for omni response when programming so you don't have to concern yourself with channel assignments). In the diagram, the 1 kHz fader is about to be edited.

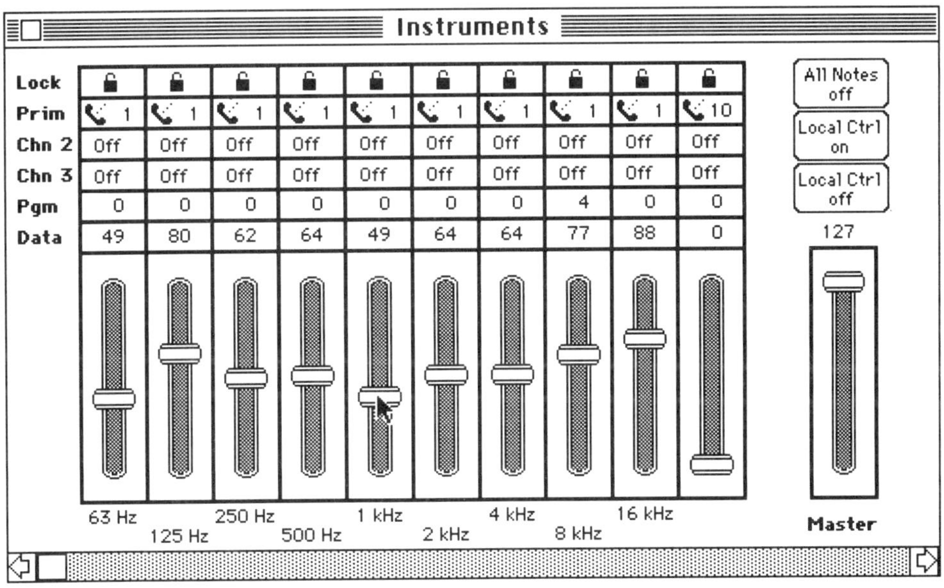

Fig. 6-7: Metro's faders set up for controlling a graphic EQ.

Fig. 6-8 illustrates the same fader functions implemented in Performer, a sequencer from Mark of the Unicorn. The fader assignments were created in the upper window; the lower window is the result of creating a fader "console." With a console, you choose whether the faders will be large or small, and oriented horizontally or vertically.

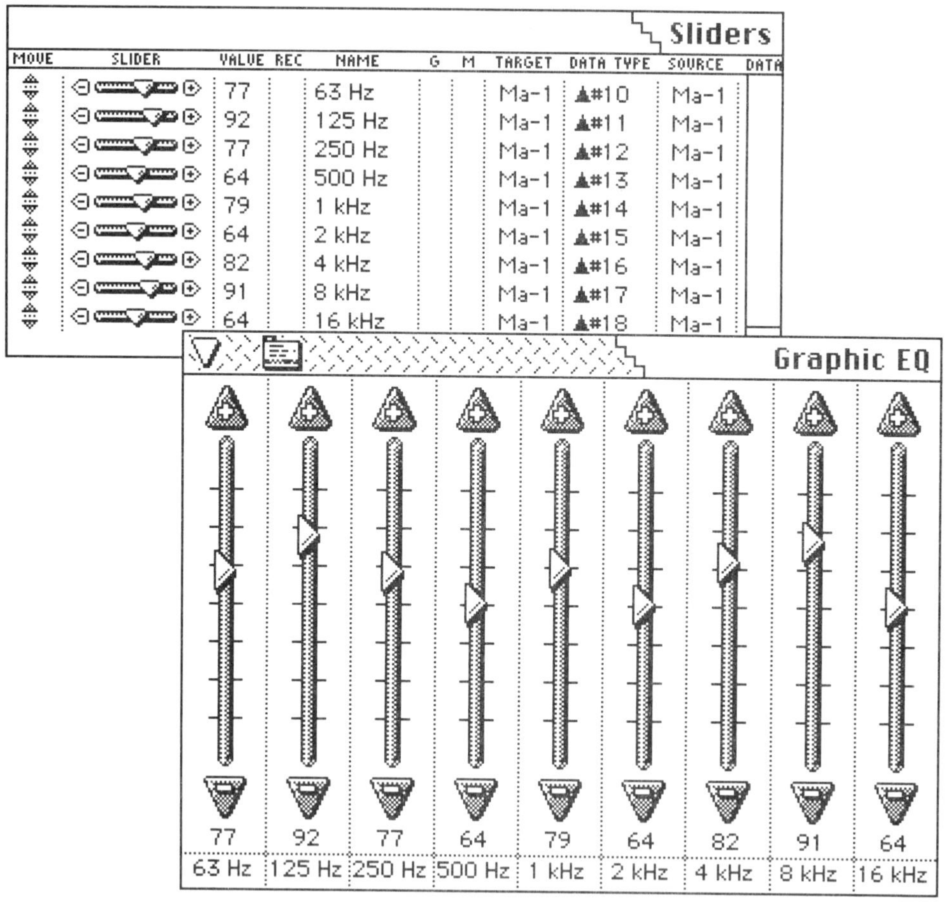

Fig. 6-8: Performer's faders set up for the same function. Note that the upper window shows the controller assigned to each fader under "data type."

Details

If you save a program whose parameters have been edited with on-screen faders, the multieffects may not store the edited settings. This is because the controller changes may simply add to or subtract from the contents in memory, rather than actually change the memorized values. If sending controller information for a specific parameter forces the LCD to that parameter, and the display also updates the parameter value as you change it, then the multieffects will probably remember your edits.

Overall, sequencers can even make signal processor programming easier. Not all signal processors are equally well-suited to this technique, but more often than not, using your sequencer's faders beats making changes from the device's front panel.

Delay-Based Drum Applications

Multieffects can produce wonderful special effects with drums. Here are some favorites; most require only mono or stereo delay.

Tuning Percussive Sounds

Delay lines can give a sense of pitch to unpitched sounds (white noise, cymbals, handclaps, etc.), thus helping to "tune" drum and percussion tracks to a specific key.

Set an initial delay time between 0 and 10 ms, and turn up feedback as high as possible short of oscillation (positive feedback gives the strongest sense of pitch). Set modulation amount to zero, and output mix to 100% delayed sound.

Use the fine delay parameter to vary the "tuning," and feedback phase to vary the timbre. To make the effect less obvious, decrease the amount of feedback and/or set the mix for more straight signal.

Automatic Tom Flammer

A drum "flam" occurs when a drummer strikes the same drum twice in quick succession. This sound resembles slapback echo, and is often applied to toms.

To flam the toms without flamming the other drums, assign the toms to their own outputs (in addition to the main stereo outputs, many drum machines let you assign particular drums to auxiliary outputs). Patch the tom outputs to the multieffects input, the multieffects output to two mixer channels, and the main drum outputs to two additional mixer channels (Fig. 6-9).

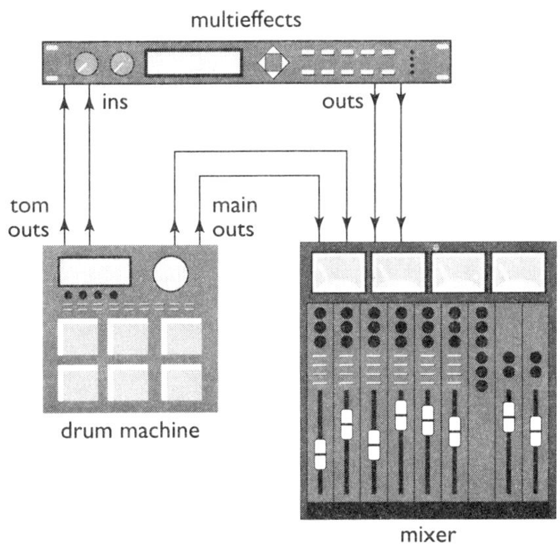

Fig. 6-9: Signal flow for automatic tom flamming.

For the initial delay, select a 25 to 50 ms flam repeat time. Since this parameter sets the time interval between the two drum sounds, it should ideally relate to the music's tempo (see "Relating Echo Times to Song Tempos" in Chapter 3). Use minimum feedback to hear a single flam instead of multiple echoes, and minimum modulation. Initially set the output mix to 50% straight and 50% delayed sound; emphasize or de-emphasize the flam by selecting more or less delayed sound respectively.

Note that you can use the bypass switch to bring the flam effect in and out as desired.

Ultra-Fat Toms

Here's an application that makes good use of pitch transposition. Take a separate submix of the toms (as in the previous application), and run them into the multieffects. Select downward pitch transposition between 1 and 5 semitones, with no feedback. Adjust the output mix for the desired blend of straight and processed toms (start with 2/3 straight, 1/3 processed). Optionally use predelay to create a "bigger and fatter" sound. The end result is toms that sound really huge.

This application also works well with handclaps and sometimes with snare.

Mixing With EQ

You don't always have to use volume to change a drum sound's level in the mix; EQ can work equally well, especially with pitched drum sounds such as toms.

Program a parametric equalizer for a bit of boost (around 3-6 dB) and moderate bandwidth (about 1/3 octave). Adjust the frequency parameter to "zoom in" on the sound you want to boost, then if necessary, increase the amount of boost and/or narrow the bandwidth.

You can also cut instead of boost to make a particular drum sound *less* prominent in the mix.

Drum Machine Handclap Realism Enhancer

Unlike humans, drum machines make exactly the same handclap sound every time—which is not always desirable. Processing the drum machine handclaps with a tight, slightly modulated echo creates a more randomized handclap effect that sounds more "humanized." Depending on how the modulation affects the echo time at any given moment, the delayed handclaps can occur further apart from, or closer to, the straight handclaps. Modulation also adds some pitch-shifting that subtly changes the handclaps' timbre.

You need to be able to send the handclaps to a separate output, which then feeds the multieffects (see the previous application, "Automatic Tom Flammer," for information on using individual outputs with a mixer). If your drum machine doesn't offer separate outputs, then try panning the handclaps to the left or right output, and pan the remaining drums to the other output.

Set the initial delay between 10 and 40 ms; 20 ms seems about optimum. Initially select minimum feedback, although adding a subtle amount might improve the effect. Set the modulation rate to around 4 Hz, and turn up the modulation amount as much as possible short of obtaining an "out-of-tune" sound. The more you increase the modulation width, the more radical the timing variations between the delayed and straight handclaps. Initially set the output mix to 50% straight and 50% delayed; for a subtler effect, reduce the delayed signal.

Extending Electronic Drum Cymbal Decay

Storing long-decay drum sounds (such as cymbals) in digital drums requires a lot of memory. To conserve memory, and therefore cost, most companies shorten these sounds by truncating their decays. Some delay can artificially extend an electronic cymbal's decay for a more natural sound.

Take the cymbal's individual output and plug it into the multieffects. The object is to add some very tight, almost reverb-like echo with a fair amount of feedback.

Set the initial delay for 23-25 ms, about 50% positive feedback, minimum modulation width, and output mix to about 65% straight and 35% delayed sound.

Increase feedback to lengthen the cymbal's "decay," and use the mix parameter to set the synthesized decay level. The delay must be short enough so that you do not hear individual echoes, but rather, a "stream" of sound.

"Humanizing" Drum Machine Parts

Producer/software engineer Michael Stewart has postulated that having different drums hit ahead of, or behind, the beat produces different types of "feel," and that these effects are quantifiable. Subsequent research has borne out his theory. As some examples, hitting a snare drum slightly later than the beat produces a loose type of groove, and gives a "bigger" sound. Hitting the snare a bit ahead of the beat gives a more rock and roll sound. Hits that occur way in front of the beat produce a more "nervous" feel, and so on.

With multieffects, it is very easy to delay a particular drum by assigning it to an individual output, delaying it by a fixed amount (typically 1 to 10 ms), and feeding the delayed drum sound into a mixer. The only way to make a drum appear ahead of the beat is to delay the sound of the entire kit *(i.e.,* delay the main output signal), but *not* delay the individual output of the drum that you want to have appear in front of the beat.

This application requires a drum machine with multiple outputs so you can process some drums but leave the rest alone. Note that if you are using the drum machine's main outputs as well as some of its individual outputs, you do not want the drum sound being processed to also appear in the main output—otherwise, you might hear flanging effects. With some drum machines, simply plugging into an external output will disconnect the drum sound from the main output. With other machines, it will be necessary to change the mix so that a given drum does not appear at the main outputs.

Make sure feedback is at minimum. However, to randomize the sound somewhat you might want to add some modulation, set to a fairly slow speed, to vary the delay over a small range. Modulation speed determines the timing difference between successive drum hits. If the speed is too slow, the timing differences between successive hits will not seem all that great.

As you play the drum part, slowly vary the delay time until the drum falls into the right "groove." Remember, this is a very subtle difference; if the drum sounds obviously late, then you probably have too much delay.

"I-Like-That-Hi-Hat" Hi-Hat Processor

This application was suggested by musician/author Craig O'Donnell, who discovered it while trying to make an uninteresting hi-hat sound interesting. The patch produces a rough, industrial-noise hi-hat by jamming as much signal as possible into the multieffects, then adding some modulation.

Turn up the input control so that the overload LED is on as much as possible—we're going for maximum nasty here. Select a short to medium delay (whatever sounds best in the mix), start with minimum feedback and increase later if desired, add some modulation, and select a slow modulation rate. Initially set the output mix for 50% straight and 50% delayed sound.

When listening to the delayed sound only, a short delay (under 5 ms) will prevent the hi-hat sound from lagging noticeably behind the beat. When listening to a mix of straight and delayed sounds, especially with longer echoes, try to synchronize the echo time to the song's tempo (see "Relating Echo Times to Song Tempos" in Chapter 3).

Varying the input level control for different amounts of distortion provides different timbres; generally, the more level, the "grittier" the sound.

Delay-Based Special Effects

Robot Voices

Sci-fi films often use a vocoder (an expensive signal processor; see Chapter 3) to create mechanical-sounding "robot voices." Although some multieffects have a vocoder module, this application can give a similar effect and requires only delay.

The settings for this patch are quite critical, so be prepared to experiment. Start with 15 to 20 ms of initial delay, as much positive feedback as possible short of runaway feedback, minimum modulation, and processed sound only (no straight sound). As you speak into the microphone, vary the delay time until your voice acquires a suitably metallic/industrial timbre.

One popular vocoder effect is to change the pitch of the vocoded signal; to imitate this effect with a delay line, vary the delay time. This is a natural for MIDI control (see Chapter 5), since you can use continuous controllers to add quick, predictable delay time changes.

"Boing" Pitch Shifting

Varying the delay time while feeding a sound into a delay effect can shift the sound's pitch for as long as the delay time is being varied. The magnitude of the pitch shift depends on the delay time; longer delays give greater pitch shifts. If you stop varying the delay time, the signal returns to normal pitch. Using delay to pitch-shift a signal gives a very different sound quality compared to using a pitch shifter module.

Delay-based pitch-shifting is particularly useful with drums and similar percussive sounds, but you can also create some pretty bizarre vocal effects. Since pitch shifting requires varying the delay time, consider using a MIDI footpedal or other form of MIDI control to vary delay.

Initially select a 25 ms delay, minimum feedback, minimum modulation, and delayed sound only. For more pronounced pitch-shifting, increase the initial delay. To accompany the pitch-shifted sound with some straight signal, set the output mix parameter for more straight sound.

Note that increasing feedback and/or altering the feedback phase can produce some pretty novel effects.

Mono-to-Stereo Conversion (Synthesized Stereo)

Suppose you want to record a stereo drum sound on tape, but have only one spare track available. Consider recording the track in mono, but then using delay during mixdown to artificially spread the mono track into stereo. This application requires a stereo delay effect (and a fair amount of tweaking to get the right sound).

Set one channel's delay time as close to 0 ms as possible, and the other channel to around 18 ms. Select minimum feedback, minimum modulation, and delayed sound only.

This is a tricky application to get right, because combining a synthesized stereo signal back into mono (as would happen if you heard the stereo signal over a mono AM radio or standard television) can sound different from the original mono signal. This problem occurs because the delayed channel synthesizes stereo by passing the straight signal through a short delay; re-combining a short-delayed signal along with a non-delayed signal can produce frequency response changes. Therefore, monitor the stereo sound panned to center (mono) as well as in stereo, and carefully adjust the delay time for the least amount of tonal change when switching between synthesized stereo and mono. Setting a longer delay time is not always the answer, as this might add a slapback echo effect.

For a more unusual sound, try adding some feedback.

Vocal With Stereo Spread Sibilance Echo

When adding echo to voice, sometimes the echo "steps on" (obscures) the main vocal. One solution is to thin out the echo sound by restricting its frequency response. Generally, cutting out the bass and echoing only the highs produces a tight, sibilance-enhancing effect that makes the vocal more prominent and intelligible. Since this patch separates the sibilance echo sound from the main vocal, you can place the echo sound anywhere in the stereo field; however, in most cases placing the straight vocal in the middle of the stereo field works best.

This requires a multieffects that lets you set up modules in parallel so that the echoed signal goes through EQ, but the straight signal does not. Fig. 6-10 shows a typical algorithm that lets you do this.

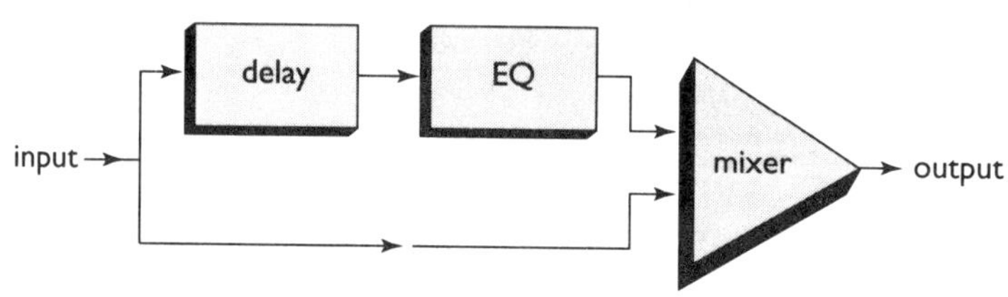

Fig. 6-10: Algorithm for adding treble to the vocal's echoes, but not the straight vocal.

Adjust the controls as you would for standard echo, except set the output mix for delayed sound only. Now thin out the echo sound by adjusting the equalizer for a treble boost, bass cut, or both—whatever works best. However, note that boosting the treble is more likely to overload the multieffects than cutting the bass, and also, you may need to re-adjust the multieffects' input level control (and possibly feedback control) for different equalizer settings.

Warped Record Simulator

Those who do sound work for films or the theater sometimes get requests for pretty strange sound effects. If you ever need to simulate a warped record, here's the patch.

Feed the sound or program material to be "warped" into the multieffects input, and call up a time delay effect. Set the initial delay in the slapback echo range (around 50 ms), select minimum feedback, and adjust the output mix for delayed sound only.

Vary the modulation width for the desired amount of warpage. The more modulation (and/or delay time), the greater the simulated amount of warp.

To simulate a warped 33-1/3 RPM record, set the modulation speed as close as possible to 0.55 Hz. To simulate a warped 45 RPM record, set the speed to 0.75 Hz. To simulate a warped 78 RPM record, set the speed to 1.3 Hz.

Short Wave Receiver Sound Effects

Intentionally feeding very high frequency signals into a multieffects can cause interference between the input signal and the multieffects' internal circuitry. This "aliasing" effect typically produces random, strange, unpredictable sounds at the audio output that resemble tuning across a short wave band. This patch may not work with high quality devices that include extensive input filtering to keep out excessively high frequency signals; however, less expensive units usually work quite well for this admittedly bizarre application.

Plug a signal generator or synthesizer (use a waveform with lots of harmonics, such as a square or sawtooth wave), call up a time-delay based patch such as delay, pitch-shifting, or reverb, and turn up the output as high as possible. It's impossible to give suggested control settings since we're doing an "illegal" effect, but try long delay times and/or large pitch transposition intervals.

If none of your efforts produce weird sounds, congratulations: you have a high-quality multieffects with excellent input filtering. Unfortunately, you will not be able to abuse this kind of delay for the desired results.

Mono to Stereo to Hard Stereo Conversion

This unusual application uses a stereo delay module to continuously change a mono signal from mono to stereo with some center channel information, to "hard" stereo (left and right only) with very little center channel information.

Set one channel's initial delay time for about 5 or 6 ms, and the other channel for 10 to 11 ms. Initially select minimum modulation, minimum feedback, and the same feedback phase for both channels.

The output mix parameter is crucial. For a mono output, set both channels for 100% straight signal. For stereo, set both channels for 50% straight and 50% delayed sounds. For hard stereo, set both channels for 100% delayed sound.

Caution: With short delays, spreading a mono source into stereo and then re-combining the stereo outputs back into mono may change the tonality; with long delays, you may hear an objectionable slapback echo effect. Therefore, select a delay setting that is short enough to avoid echo effects, but long enough so that the signal sounds acceptable when mixed back into mono.

Psycho-Acoustic Panning

This application sounds outstanding with held chords, drones, sustained guitar, complex drum machine patterns, and the like. It is particularly effective when wearing headphones, since the sound pans and swirls inside your head. (The "panning" is caused by psycho-acoustic phase and delay interactions rather than standard level changes.) This application requires a stereo delay module with variable modulation for each channel.

The following settings are a point of departure; be sure to experiment.

Set one channel's initial delay to about 6 ms, and the other to about 12 ms. Initially add about 20% to 50% feedback (negative phase is preferred, as it gives a somewhat "gentler" sound than positive feedback). Add enough modulation to create an obvious effect, but not so much as to sound "out-of-tune."

The modulation speed adjusts the panning rate. Set one delay's speed control to approximately 1 Hz, then offset the other delay's speed slightly from 1 Hz. Program both delays for 100% delayed sound.

The most important controls for this application are modulation width, modulation speed, and feedback. Increasing modulation width makes the panning more pronounced, but will also make the sweep effect less even-sounding. Add as much modulation as possible, consistent with a good-sounding sweep.

The speed controls should be close to the same rate but not synchronized. If the panning effect is not apparent enough, increase the speed. Finally, feedback should be kept low enough to avoid an overly "metallic" tonality.

Getting Nasty

One of the complaints about multieffects, especially from those who play in industrial or hardcore bands, is that these processors are too "clean" and seem tailored to musicians who want to make "popular," contemporary sounds. But that doesn't have to be the case.

Although most multieffects are designed with normalcy in mind, it is possible to abuse multieffects (and other contemporary devices). In fact, digital distortion is so incredibly ugly there is talk of outlawing it in some states (just kidding).

The key is to look for devices with lots of programming options. For example, if you can change the order of effects in a multieffects unit, you can do weird things such as put reverb or pitch shifter through distortion, and then overload an equalizer. Devices with effects algorithms, where specific effects are put in specific orders, are easier to program but don't give as much latitude.

MIDI continuous controllers are another way to get ugly sounds. Find the parameters that glitch when you change them, and sweep the parameters back and forth with a pedal as you play. Or, use a small hardware sequencer to generate bizarre controller variations. You don't have to sync it to anything if you're going for Maximum Ugly.

Although you can always scour second-hand stores for older, messed-up gear, one advantage of digital ugliness is that it's predictable and doesn't depend on something like a part that's starting to fail. As someone who on occasion likes to spray out sounds that empty rooms and frighten small animals, I've found that some digital multieffects offer enough options to make truly weird—as well as truly beautiful—sounds.

CHAPTER 7

Troubleshooting (and Updating) Your Multieffects

Although multieffects devices tend to be reliable, problems can occur. The tough ones are best left to the experts, but in some cases, you may be able to fix the problem yourself and save a trip to the repair shop.

Rules of Successful Troubleshooting

- Begin by taking no action at all: just think about the problem logically. Gather as much data as possible, since collecting this data will often isolate the problem. For example, if a multieffects worked at your house yesterday but doesn't work at a friend's house today, odds are the multieffects is not the problem, but something that changed—such as a cable, mixer, or amp.

- Maintain a positive attitude about being able to fix the box, since you will think more clearly if you're not agitated. Don't get angry or panic; often the problem is nothing more complex than a misset control or incorrectly adjusted parameter.

- Don't be afraid to take a box apart in order to see what's going on inside—but also observe common sense precautions, such as *never* working on AC powered equipment while it's plugged in. You can access the innards of many rack mount devices by simply removing a few screws that hold the top and bottom panels in place. Be careful, though; if it's not easy to open up the box, don't.

- Visually inspect the effect. See if any jacks are loose or don't feel "tight" when you plug into them, if the AC power cord looks damaged, if one switch doesn't feel as solid as the others, and the like.

- Don't get in over your head. If you see something that's obviously wrong, by all means fix it. However, it's always a drag to create a new problem while trying to solve an old one, so recognize that in some cases you'll need a qualified technician to help you out.

- Remember that most problems are minor problems or due to human error, so be on the lookout for the obvious, not the obscure.

- Read the manual!

Specific Cures for Specific Problems

- **Totally dead unit.** AC powered equipment usually includes a fuse. Fuses are the musician's security guard, as they protect circuits from damage under an overload condition by interrupting the flow of electricity into the device. When a fuse blows, nothing happens—no lights, no transformer hum, nothing.

  When a piece of equipment blows a fuse, it's logical to assume that there's some kind of serious problem. However, this is not always the case. Mechanical shock, old age, a fluke power surge, and similar problems can wipe out a fuse.

  If the fuse is easily accessible, hold it up to the light and see if the fuse element is blown *(i.e.,* there's a gap in the element—see Fig. 7-1). However, the fuse may be mounted internally to the unit, in which case you'll need to do some disassembly before you can fix the problem. *Never* replace a fuse with one of a higher rating, and if there are any indications of suspicious behavior (smoke, burning smells, sparking, hot parts), shut the power off immediately—odds are you have a serious problem that requires professional attention.

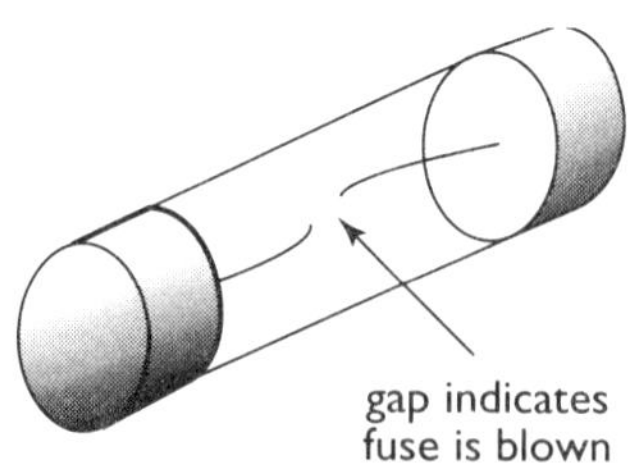

Fig. 7-1: Checking to see whether a fuse is blown.

- **Misset parameters and "obvious" mistakes.** Many so-called "broken" units are often the victims of a misset parameter. For example, an effect which appears to have no output signal may simply have the output level parameter set to zero. Other "obvious" problems include reversing the audio input and outputs connections, not plugging plugs all the way into their jacks (another common problem), forgetting to plug in the AC power cord, and so on. Always look for simple problems such as these before taking the box apart and proceeding to the next level of troubleshooting.

- **Internal battery failure.** Programmable devices have an internal battery to power the memory and keep the programs alive while the device is turned off. If your multieffects loses all the user programs, it's possible that the battery has gone to battery heaven. Replacing it may require a trip to the repair shop, as the battery will often be soldered in place; overheating a battery during the soldering process could cause it to explode. In any case, hopefully you've backed up all your programs so you can reload them once you replace the battery.

- **Lockups and re-intialization.** If your multieffects behaves strangely (such as "freezing" on one display) there may be a software glitch. Turn the unit off, wait 30 seconds or so, then turn it back on. If this doesn't help, most devices have some sort of reset procedure (also called re-initialization), such as holding down specific buttons while turning on power, that causes the unit to return to its initial state. Check the manual for details on this procedure, or call the manufacturer. Resetting will often erase any user programs, which is yet another argument for being meticulous about backing up your work.

- **Loose internal connectors.** Many multieffects have ribbon connectors and other cables that can come loose if you're on the road a lot. Even if the cables haven't become disconnected, you can sometimes solve problems by *gently* jiggling each connector in its matching socket.

- **Oxidized integrated circuit pins and sockets.** If an integrated circuit (IC) is mounted in a socket, push down *extremely* gently on the IC to make sure it's well-seated in the socket. This can also scrape off any oxidation if present.

- **Loose or bent jack.** A loose jack can cause problems ranging from snaps, crackles, and pops to complete non-functionality. Jiggle the input and output jacks to see if they move; if they do, tighten the mounting nut. However, be careful not to overtighten. Sometimes the jacks mount directly to the unit's circuit board; overtightening may stress the board and lead to a *very* expensive repair job.

- **Intermittent controls.** If a volume control or data-setting control acts intermittently, dirt or dust may be caught in the control. Go to your local electronics store and purchase a spray can of contact cleaner that is specifically indicated as being safe for plastics. Unless the control is sealed (in which case dirt probably didn't get into it anyway), there will be some kind of opening into which you can spray the contact cleaner. Typically, this will be a space near the control terminals that solder to the circuit board or wires (Fig. 7-2). Spray some contact cleaner into this space (direct it toward the control's center), and turn the control's knob back and forth vigorously. This wiping action distributes the contact cleaner evenly. Should a second application fail to solve the problem, the control needs to be replaced with one of the same value.

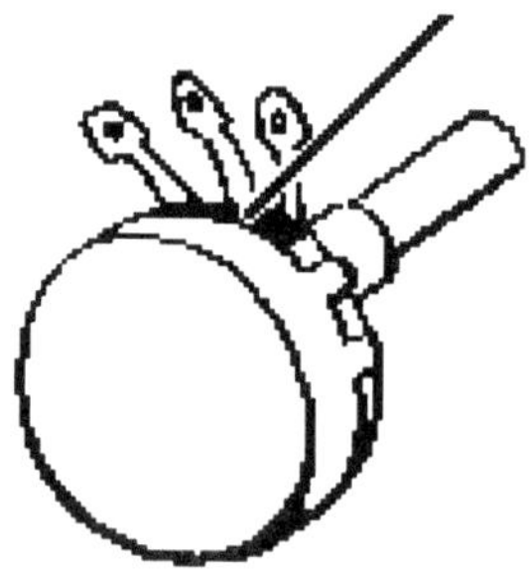

Fig. 7-2: Spraying contact cleaner into a control or switch can sometimes prevent intermittent operation.

Do it Yourself Repairs vs. Warranties

You may notice that we've covered all the above subjects without once mentioning that attempting a do-it-yourself repair may void your warranty. Well, now we've mentioned it!

Updating Your Multieffects

Because multieffects are software-based, they can often be updated by replacing the chip (integrated circuit) that contains the software. Updating occurs for two reasons: either because bugs were discovered that need to be fixed, or the manufacturer wants to add new features to remain competitive with newer units. This is another reason why it's so important to send in your warranty card—manufacturers often do mailings to let customers know about updates.

Although a chip update is an easy way to gain new features, if you don't do it right you could end up with anything from temporary frustration to a massive repair bill. While there's no need to be intimidated, you can't be casual about the process either. If you want a successful chip transplant, it's good practice to follow these 10 steps.

1. **Be prepared.** Save all memory contents of the unit being updated since after updating, it's often necessary to re-initialize. Then unplug the device, find a well-lit work space, and gather your tools:

   - Screwdrivers (for disassembling the case to get to the chip)
   - IC inserter/extractor (Radio Shack #276-1581; see Fig. 7-3)
   - Needlenose pliers or IC pin aligner (Radio Shack #276-1594)
   - Conductive ground strap (Radio Shack #276-2397)
   - Small piece of aluminum foil
   - Paper and pencil

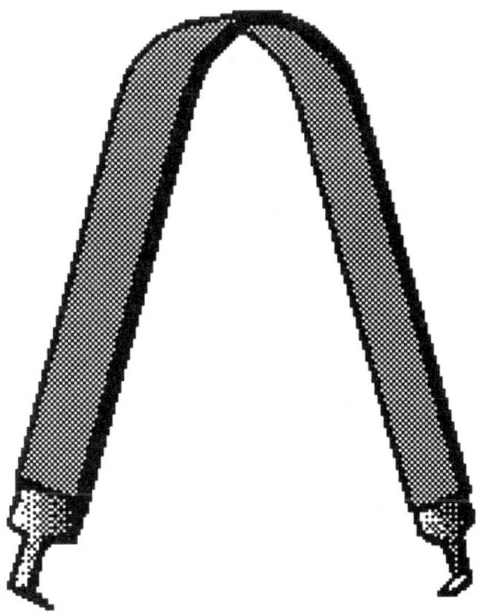

Fig. 7-3: Inexpensive IC extractor. Squeezing on the spring forces the clips together, which inserts between the chip and socket at the top and bottom of the chip. Rocking and pulling gently upward lifts the IC out of its socket.

2. **Locate the chip to be replaced.** This may or may not be easy. Some units have a "trap door" that escorts you right to the machine's innards, where you can easily replace the chip. With other devices, you may have to disassemble the case, remove circuit boards, and/or unbundle cables to get at the chip. Double-check that the chip designation is the same as the replacement IC.

   Remember that you are dealing with a fragile piece of gear where one mistake can cause serious problems. If the chip's location is not obvious, you're probably better off having an authorized service center do the update for you.

3. **Write down the chip orientation.** A chip will have a notch or dot at one end (see Fig. 7-4), and it is vital that the replacement chip be oriented in the same way.

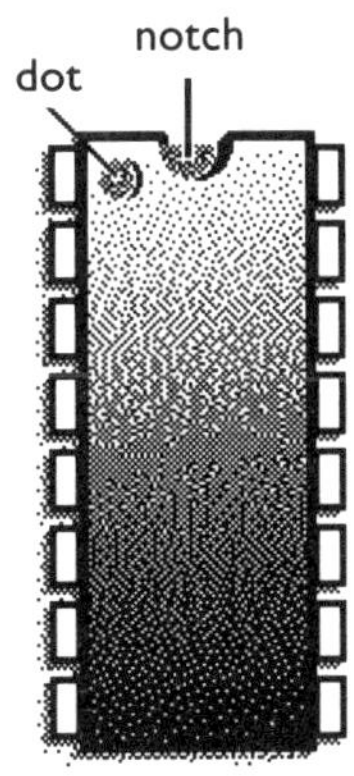

Fig. 7-4: A typical integrated circuit. The dot or notch indicates the "top" of the IC (i.e., pin 1 is in the upper left-hand corner).

Failure to do so will probably fry the chip, and may damage the power supply too. Also, sometimes two or more chips will need to be replaced. Write down their locations and any distinctive markings or part numbers. Do not trust your memory! If the phone rings in the middle of a chip change or you have to run off to a session, you may not remember which chip went where when you return.

4. **Discharge yourself of static electricity.** Touching circuitry after you've accumulated a static charge *(e.g.,* from walking across a rug on a day with low humidity) could destroy sensitive silicon parts. Worse yet, sometimes static damage weakens the chip instead of outright killing it, causing intermittent problems (a repair person's nightmare) or a complete failure when you least expect it.

   To discharge yourself, place one end of the conductive ground strap around your wrist, and attach the other end to the multieffects' chassis ground. If you don't use a ground strap, at least touch a metal ground to discharge any static electricity from yourself before handling any chips, and don't do anything that could accumulate a charge while you're replacing the chip.

   Having said all that, modern chips are remarkably resistant to static damage—but there's no need to tempt fate.

5. **Remove the old chip(s).** If the chip is soldered in, forget it. Close the unit back up and go to a service center. Otherwise, use the IC puller to remove the chip. Pull straight up and out of the socket; you may need to rock back and forth *very* slightly to loosen the grip of the socket on the chip, but avoid bending the chip's pins as you pull the chip out. Place the chip on the piece of aluminum foil so that the pins are contacting metal.

   Since you're replacing the old chip(s), why worry about treating it with care? Simple: The new chip may be defective, or there may be bugs in the new software that make you want to go back to the original chip.

6. **Insert the new chip(s).** While you're still connected to the ground strap, remove the new chips from their protective foam, foil, or plastic IC carrier. If the IC pins aren't straight, use an IC pin aligner or needlenose pliers; unstraightened pins can bend under the chip when inserted, or worse yet, break off. Insert the chip in the IC insertion tool, then plug the chip into its accompanying socket. Remember to double-check the orientation of the notch or dot.

   Incidentally, you don't absolutely *need* an IC insertion tool; you can always line up one row of pins in the socket, then push gently against the opposite side of the chip until the other row of pins lines up with the other side of the socket. But the proper tool can cost a lot less than a botched IC insertion.

7. **Push down gently on the chip.** Apply even pressure at both ends of the chip to make sure it's well seated. Usually there will be a little resistance as the chip seats firmly in the socket, but *be careful* not to push too hard if the circuit board under the chip isn't well supported—bending the board could break a trace or solder connection, leading to a big-time repair bill.

8. **Double-check the chip orientation against your original drawing.** Once you're sure it's okay, close up the unit.

9. **Re-initialize the device.** This may not be required, but is good practice anyway. If you do re-initialize, then you'll probably want to reload the memory contents you saved back in step 1.

   Congratulations! You're updated.

10. **Return the old chip.** This is the *coup de grâce:* after you're sure that everything checks out okay and you're satisfied with the new software, return the old chip to the manufacturer (sometimes these can be reused internally for prototyping). Who knows, the service people may be so impressed by your level of consideration that they'll give you special treatment if you ever need a quick repair job.

Glossary

Please note that most of the following terms are defined within the context of multieffects. For example, under the definition for program, we talk only about the program in a multieffects, not the act of creating software for computers.

Active Describes a circuit or device that requires a source of power. Example: Active guitar pickups include an amplifier, and require a battery for operation.

Algorithm (1) A combination of effects, arranged in a particular order, that creates a particular type of sound. (2) A routine a computer invokes to accomplish a task.

Ambience A room's acoustical characteristics.

Amplify To increase a signal's level.

Amplitude A term describing signal level. Example: A loud stereo puts out a high amplitude signal.

Analog Continuous in nature. Example: A light switch exemplifies digital operation, since there are only two states—on and off. A light dimmer exemplifies analog operation because the light can vary continuously from full off to full on.

Attack delay A signal processor that synthesizes a volume fade in, and is generally triggered by each new note or chord.

Attack time The time required for a circuit or software routine to react to a trigger or event.

Attenuate To lower a signal's level. Attenuation is the opposite of amplification.

Automatic Double Tracking (ADT) A time-based signal processor that simulates the effect of playing a part, then overdubbing a second part to give a thicker sound.

Auto pan To change a signal's spatial position in the stereo field via some modulation source.

Balanced line A method of interconnecting audio equipment using three conductors: hot, neutral, and ground. Balanced lines tend to reject hum and radio frequency interference; they can also run long distances without excessive signal deterioration. Most multieffects use unbalanced inputs and outputs for compatibility with standard guitars and amps. However, some devices offer balanced line inputs and/or outputs for interfacing with professional studio equipment.

Bandpass filter A signal processor that boosts only those frequencies around its resonant frequency, while rejecting higher and lower frequencies.

Bandwidth A device's frequency response, or the *width* of the *band* of signals that the device passes. The bandwidth of the human ear is approximately 10 octaves (20 Hz to 20 kHz). Bandwidth also refers to the range of signals passed by a filter.

Bank select message A MIDI message that allows for choosing 16,384 banks of 128 programs. This circumvents MIDI's original 128 program limit.

BPM Beats per minute. A measure of tempo; the number of quarter notes that occur during a minute of music.

Bypass switch Allows comparing processed and unprocessed sounds by switching an effect in and out of the signal path.

Chip Slang for integrated circuit. See integrated circuit.

Chorus A signal processor that processes a single instrument to mimic the sound of two instruments playing *en ensemble.*

Clipping indicator A light that indicates the onset of distortion. Signal levels should usually be adjusted so that the clipping indicator lights rarely, if at all.

Compression ratio The ratio of a compressor's output level change compared to a given input level change. Example: If increasing the input signal level by 6 dB yields an output increase of 2 dB, the compressor has a 3:1 compression ratio.

Compressor A signal processor that evens out variations in dynamic range by amplifying soft signals to make them louder, and attenuating loud signals to make them softer.

Compressor/limiter A signal processor where compression begins only above a certain threshold.

Continuous controller message A MIDI message, generated by a MIDI control device (*e.g.*, MIDI footpedal or sequencer), that can alter multieffects parameter values in real time.

Daisy-chain To interconnect equipment so that each output feeds no more than one input.

Decay time The time for a function to return to its quiescent state after the end of an event.

Decibel A unit of measurement that defines the ratio between two audio signals.

Default A value assumed initially until changed. Example: A multieffects might default to program 01 when first turned on, or to the last program selected.

Digital Having one of two possible states (on or off, yes or no, etc.). Digital circuitry performs logical decisions based on a series of yes/no questions. Example: Is a button being pressed? If yes, an action needs to be taken.

Digitization The process of converting an analog (continuously varying) signal into a digital (series of numbers) signal. Example: A mercury thermometer is an example of analog temperature measurement. A digital thermometer digitizes the temperature into a specific number, such as 98°.

DIN connector The type of connector used for MIDI signals. Although DIN connectors come in several different types, MIDI uses 5 pin, 180° connectors.

Distortion Added components of a signal that were not part of the original signal.

DSP Digital Signal Processor. A special-purpose computer chip designed for processing digital audio.

Dynamic range The difference between the softest and loudest signals a device can handle.

Early reflections The first group of echoes that occur in an acoustical space; you hear these as discrete events, as opposed to the subsequent "wash" of room reflections.

Effects loop Allows plugging an external effect or group of effects into a multieffects or amp. This accommodates some "special" effect (such as a vintage wa-wa pedal) that's not included in the multieffects so that it can be a part of the signal chain.

Envelope follower A modulation source that tracks the dynamics of an audio input signal.

Equalization (EQ) The process of altering a signal's tone by changing the frequency response of the device through which the signal passes.

Equalizer A signal processor that emphasizes or de-emphasizes certain frequencies in order to change a signal's timbre.

Exciter A signal processor that enhances a signal's brightness and adds "airiness" without the use of EQ.

Feedback To send some of an effect's output signal back to the input.

Filter A circuit or software routine that alters frequency response. Example: A lowpass filter passes all signals lower than a certain frequency, whereas a highpass filter passes all signals higher than a certain frequency.

First reflection The first discrete echo that occurs in an acoustical space.

Flanger A time-based signal processor that imparts a whooshing, "jet airplane" sound.

Glitch An unexpected, and usually upsetting, phenomenon.

Global Having the quality of affecting the entire multieffects, rather than simply individual programs.

Graphic equalizer A signal processor that uses multiple bandpass filters to split the audio spectrum up into several bands, with an individual boost/cut control for each band.

Hard clipping The condition where an output signal remains undistorted up to a certain point (the clipping point), then becomes extremely distorted as the input increases past that point.

Hertz (Hz) A unit of measurement for cycles per second. Example: AC wall current oscillates at 60 Hz.

Highpass filter A signal processor that passes signals above a specified cutoff frequency, and rejects signals below the cutoff frequency.

Initialize To return a device to its original state, prior to any editing or programming.

Integrated circuit A tiny electronic device that crams dozens, hundreds, or sometimes even thousands of components on a piece of material called a "chip" (made of silicon), which is typically about the same size as one of the letters in this sentence. These components work together to form a particular type of electronic circuit such as an amplifier, computer memory, etc. Since chips are so small, they're packaged in some form of carrier that includes pins for easy connection to external components.

Intelligent harmonization The ability of a pitch transposer to create harmony lines based on rules of harmony, rather than being limited to simple parallel harmonies.

Inversion As applied to continuous controllers, this sets whether increasing controller values increase (+ scaling) or decrease (- scaling) the target parameter value.

Kilohertz Abbreviated kHz. A unit of measurement of frequency that equals one thousand cycles per second.

LCD Liquid Crystal Display. A type of display technology characterized by high visibility, low cost, and low power consumption.

LED Light Emitting Diode. A solid state light. LEDs are most commonly available in red, although yellow, orange, green, blue, and dual color (bicolor) LEDs are also available.

LED meter A hardware device that measures signal level, consisting typically of 4 to 20 LEDs arranged in a column or row. Sometimes different colored LEDs differentiate between different signal conditions *(e.g.,* red is overload, yellow indicates close to overload, and green indicates normal operation).

LFO Low Frequency Oscillator, a circuit or software routine that provides a cyclical (periodic) control change for modulating a parameter. Also see modulation.

Limiter A signal processor that restricts dynamic range by not letting a signal's amplitude exceed a certain level specified by a threshold parameter; signals below the threshold remain unaltered.

Lowpass filter A signal processor that passes signals below a specified cutoff frequency, and rejects signals above the cutoff frequency.

MIDI Musical Instrument Digital Interface. A communications protocol that allows musical instruments, computers, and other MIDI-equipped audio devices to communicate musically-related data to each other.

MIDI channel MIDI data can be stamped with a channel ID from 1 to 16. Each number represents a different channel, which allows up to 16 different channels of control signals to travel down a single cable.

MIDI clock message A message emitted 24 times per quarter note that provides synchronization within a MIDI system. Clock messages take priority over all other MIDI messages, and can even occur in the middle of a MIDI message.

MIDI in jack A connector that receives MIDI data.

MIDI out jack A connector that transmits MIDI data.

MIDI thru jack A connector that carries a copy of the data appearing at the MIDI in jack.

Microsecond (µs) 1/1,000,000th of a second.

Millisecond (ms) 1/1,000th of a second.

Modulation The process of altering a parameter, usually through some automatic or programmed means such as an LFO. Example: A cyclical amplitude modulation creates tremolo. A cyclical frequency modulation causes vibrato.

Modulation depth The difference between minimum and maximum modulation amounts. Also called modulation intensity.

Modulation rate A measurement of how fast modulation variations occur. Also called modulation speed.

Multieffects A signal processor containing several different effects in a single package. Most multieffects are programmable and based on digital technology.

Noise gate A signal processor that prevents audio from reaching its output when the input signal level passes below a user-settable threshold.

Notch filter A signal processor that rejects frequencies around the notch frequency, but passes frequencies above and below the notch frequency.

Omni mode A MIDI reception mode in which the receiving device accepts data appearing on any of MIDI's 16 channels.

Pan To place a signal at a specific location within the stereo field.

Parallel effects An effects configuration that splits the signal into two legs, each of which goes to its own effect. The effect outputs typically feed a mixer that recombines their signals, and possibly pans them as well.

Parameter A single, usually variable, element of an effect. Example: Parameters for an echo effect might include echo time, echo feedback, and echo/straight mix.

Parametric equalizer A signal processor that can boost or cut over a user-selectable range of frequencies, and offers variable bandwidth.

Patch Another word for program; left over from the days of analog synthesizers. Also, the process of interconnecting various devices.

Phone jack/plug Also known as 1/4″ phone jack/plug. The type of connector used at guitar outputs and amp inputs, and at the end of standard patch cords.

Pitch bend A MIDI message that indicates how much a note is being bent sharp or flat compared to concert pitch.

Pitch transposer A signal processor that synthesizes a harmony line from the input signal.

Plate reverb A large, relatively costly electro-mechanical reverberation system used in studios.

Poly mode A MIDI reception mode in which the receiving device accepts data appearing on only one of MIDI's 16 channels.

Predelay The amount of time delay that transpires before the onset of an event.

Preset Another term for program.

Pressure A MIDI message that indicates how much pressure is being applied to a keyboard key after it is in the key-down position. There are two types: mono (channel) pressure represents an average of the keys being held down, while polyphonic pressure produces a separate message for each key that is held down.

Program With multieffects, a combination of parameter values that gives a specific sound or effect and can be stored in memory.

Program change command A MIDI command that can call up a particular program in a multieffects.

Program change map A MIDI-related function that lets you create a table of which program should be selected in response to an incoming program change command. This is useful if you want, for example, program change 34 to call up a program other than 34.

Program material A collection of complex sounds, such as a CD, tape, radio broadcast, etc.

Programmable Having editable parameters whose values can be stored in memory and recalled as desired.

Programming The act of editing the parameters in a program.

Pseudo-parametric equalizer See quasi-parametric equalizer.

Q Another term for resonance.

Quantize To subdivide something continuous into discrete steps. Example: A digital clock quantizes time into hours, minutes, and seconds.

Quasi-parametric equalizer A parametric equalizer that includes frequency and boost/cut controls but no bandwidth control. Also called pseudo-parametric.

RAM Random Access Memory. Memory that can be overwritten and edited. User programs are stored in RAM.

Random wave A waveform whose amplitude and frequency vary randomly within maximum and minimum values, and that has generally sharp, "stair-step" transitions between different levels. Often used as a modulation source.

Recirculation See feedback.

Regeneration See feedback.

Re-initialize See initialize.

Release time The decay time that occurs after releasing a keyboard key. Also see decay time.

Reset See initialize.

Resonance The frequency at which a filter's response peaks is its resonant frequency; resonance indicates the degree of this peak.

Reverberation A signal processor that simulates the sound of playing in a large hall, auditorium, or other acoustic space.

Ring modulation A signal processor that modulates the input signal with another audio signal. Technically, the resulting signal is the sum and difference of the two input signals, with the original signal components suppressed.

ROM Read Only Memory. Memory that cannot be overwritten or altered.

ROM programs Programs that are permanently in memory and cannot be erased.

Sample An instantaneous measurement of an analog signal.

Sampling rate The number of samples taken over a period of time by a device that converts an analog voltage into a series of discrete voltage values.

Scaling As applied to continuous controllers, this determines how far a parameter will vary from the programmed setting in response to a given amount of controller change.

Sequencer A hardware device or computer-based software program that records MIDI data (including program changes and continuous controllers) in memory for later playback.

Series effects Effects that string together one after another. In other words the instrument plugs into the input of effect 1, effect 1's output plugs into effect 2's input, effect 2's output plugs into effect 3's input, and so on.

Sine wave A waveform that resembles the triangle wave, but varies in a more rounded manner rather than a linear one. Often used as a modulation source.

Slapback echo A short, tight echo sound created by delaying a signal by 25 to 50 ms and combining the delayed signal with the original signal.

Smoothed random wave A waveform that varies randomly but has no sharp transitions between different levels. Often used as a modulation source.

Soft clipping With soft clipping, the output signal becomes progressively more distorted as the input signal level increases, but in a smooth manner rather than distorting abruptly past the clip point.

Software MIDI thru A MIDI function that turns a device's MIDI out jack into a MIDI thru jack.

Speaker emulator A signal processor designed to re-create the effect of running a signal through a guitar amplifier cabinet.

Spring reverb An inexpensive, compact type of electro-mechanical reverb unit typically used in guitar amps.

Square wave A waveform that switches cyclically from a maximum to minimum value. Often used as a modulation source.

Synchronization To cause rhythmically-oriented devices to work together so that they remain locked to a particular rhythm or time base.

System exclusive messages MIDI messages that encode manufacturer-specific data in such a way that it can be sent as part of the MIDI data stream.

Tapped delay A collection of several delay lines, each with adjustable delay time and feedback. Some also include individual modulation for the different taps.

Threshold The level above or below which a signal will be triggered, compressed, limited, gated, or otherwise processed.

Through-zero flanging A type of flanging where the time difference between two signals can reach 0 seconds.

Tremolo A periodic amplitude variation.

Triangle wave A waveform that varies smoothly and in a linear fashion from a maximum to minimum value in a cyclical manner. Often used as a modulation source.

Unbalanced line A method of interconnecting audio equipment using two conductors: hot and ground. Most multieffects use unbalanced inputs and outputs for compatibility with standard guitars and amps.

Velocity A MIDI message that indicates the dynamics of a musician playing a MIDI controller (keyboard, MIDI guitar, MIDI drum pads, etc.).

Vibrato A cyclical pitch variation.

Volt A unit of measurement for electrical energy.

Wa-wa pedal A foot-controlled signal processor containing a bandpass filter with variable resonant frequency. Moving the pedal back and forth changes the bandpass filter frequency.

Wall wart Slang term for a transformer that plugs into an AC wall outlet or barrier strip.

Waveform The graphic representation of a sound.

XLR connector 3-pin connectors designed to mate with balanced inputs and outputs.

About the Author

On this 10th birthday, Craig Anderton received a guitar from his parents and a transistor radio kit from his grandmother, and since then has split his life between the technical and artistic. He recorded three albums with the 60s group Mandrake, produced three albums by classical guitarist Linda Cohen, did session work in New York, played on/mixed recordings by new age artists David Arkenstone and Spencer Brewer, and released a solo instrumental album *(Forward Motion,* distributed by MCA) in 1989.

A prolific author who is seldom seen without a notebook computer, Craig has authored articles for magazines such as *Guitar Player, Keyboard, EQ, Rolling Stone,* and *Byte.* He coined the term "electronic musician" and edited the magazine bearing that name for the first five years of its existence. He has written numerous "classic" books on musical electronics, including *Electronic Projects for Musicians, The Electronic Musician's Dictionary,* and *Home Recording for Musicians.* He also co-authored *Digital Projects for Musicians.*

In addition to serving as Consulting Editor to *Guitar Player* magazine and Technology Editor for *EQ,* lately Craig is much in demand as a lecturer—work that has taken him to over 22 states and 6 countries. He also consults to manufacturers in the music business and is responsible for some of the sounds you hear coming out of various instruments, as well as some of their design features.